CONTENTS

D1454839

1 RESPONSIBILITIES

Some of the requirements of persons doing electrical wiring are as follows:

THE ELECTRICAL CONTRACTOR

RESTRICTED CERTIFICATE OF QUALIFICATION

The Electrical Contractor must:

— Obtain permits for all of his work.

— Do all work according to code.

— Not work under a permit issued to the owner or another Contractor.

— Not take out a permit for work done by someone else.

— Not hire uncertified men on a job basis to work under his certificate. (these men are then really contracting without proper Certification).

— Not undertake to do work which exceeds his certification.

— Make sure that all electrical equipment is properly certified before it is connected. See below for a list of equipment certification marks acceptable in the province of Ontario.

— Declare in writing to the Electrical Inspector when the work is ready for inspection.

— Not cover any wiring until it is approved for covering by an Electrical Inspector, Rule 2-004(7).

THE STUDENT

The student who will familiarize himself with the information contained in this book will find it easier to write the official Government examination for a Restricted Certificate of Qualification. Try the sample examination located on page 123.

The Government Examiner will supply you with an official Code book, a set of Bulletins a copy of the Regulations, and all the paper you will need to write the exam. Do not bring your own. Just bring a pen, pencil, a calculator and a clear thinking head.

The student should understand:

— He must be at least 16 years old before he may apply for apprenticeship training; and

— That if he is applying directly for the examination without apprenticeship training, must show documentary proof that he has actual on the job experience in electrical construction for a period longer than the normal apprenticeship training. For a Branch II Certificate of Qualification, the apprenticeship training program requires four periods of 1800 hours each. Total time required is 4 x 1800 = 7200 hours of classroom and on the job training. The holder of a Branch II Certificate may do any electrical work in a single family residence or any building containing six or fewer residential units. He may also do all the wiring, including maintenance, in farm buildings; and

— That if he has successfully passed the official examination he is issued a Certificate of Qualification. This means he can now operate as an Electrical Contractor, but each city or municipality he works in will want to sell him a Contractors License — for a fee of course.

— One more thing - after hours of study and agony over the exam, he must face the Inspector in the field with each installation he does. If his work is good and his cooperation is good, he will have no problem. If it is not good, the Board of Examiners may want to talk to him about it - may even want to give him a vacation (suspend his Certificate for a period of time).

— After two failures, it appears the thing to do is to become interested in some other trade, plumbing maybe.

THE HOMEOWNER

The homeowner must:

— Obtain an electrical permit for all electrical work he does, including the wiring for ranges, dryers, furnace, extra outlets, etc.

The rules require an electrical permit for any additional wiring, any alteration, and any new wiring.

It is like a building permit; it permits the holder to install certain electrical wiring and equipment in his own home. This permit must be obtained before any electrical work is begun.

— Do all the work himself — he may not allow anyone to do any of the work covered by the permit issued to him.

An owner may do the wiring in his own home only. This is the basis on which a permit is issued to an owner. He may not install the wiring, any wiring, in his own building if it is to be rented or sold. He may obtain an electrical permit only for his own personal dwelling. He must do all the work himself. No one, other than his immediate family, such as a father, brother, son, may do any of the electrical work under this permit. It has nothing to do with payment for work done. Even an Electrical Contractor could not work under a permit issued to an owner. Treat it as you would your drivers license.

— Do all the work according to the electrical code.

— Notify, in writing, the Electrical Inspector when the work is ready for inspection. He must not request an inspection before he is, in fact, ready. The Inspector may charge a re-inspection permit fee.

— Complete the entire installation including all fixtures, switches, plates, etc. before he may have a final certificate of approval.

2 - C.S.A. CERTIFICATION - Rule 2-024 & Ontario Bulletin 2-7-0.

CSA is no longer the only testing and labeling agency acceptable in Ontario. There are now six certification agencies. Each of these agencies performs prescribed tests to insure the electrical equipment and appliances sold in Ontario are manufactured according to a rigid set of standards. Following is a list of the certification marks used by these agencies.

 CGA, Canadian Gas Association. This is what to look for on gas appliances such as furnaces, dryers, cooking stoves etc.

 CSA Canadian Standards Association. This is the old reliable and very familiar certification label. Until recently this, and the Ontario provincial label, were the only two labels acceptable in this province. This label is still acceptable just as it always was on almost everything electrical.

 ETL Inchcape Testing Service. This label is not as well known here in Canada but we can expect to see it on an ever increasing number of electrical devices.
Note the small "c" outside and to the left of the circle. This means that equipment with this label is acceptable for use in Canada. If an electrical device has this label look for the small "c". If it is not there the device is **not acceptable for use in Canada.**

 Warnock Hersey. This certification mark already appears on heating equipment. We can expect to find it on a broad range of electrical devices in the future.

 UL Underwriters Laboratories. This is a major testing agency in the United States.
Note the small "c" outside and to the left of the circle. This means that equipment with this label is acceptable for use in Canada. If an electrical device has this label look for the small "c". If it is not there the device is acceptable for use in the United States **but it is not acceptable for use in Canada.**

 ULC Underwriters Laboratories of Canada. This certification marking is used mainly on fire protection equipment. You will find it on smoke alarms used in single family homes.

Ontario Hydro marking. This marking is applied to electrical equipment which for one reason or another must be tested for acceptance in the field. This would include specialty items which have been brought in as settlers effects or items imported on a one time basis. This label indicates the equipment has been examined and found to be acceptable for use in Ontario.

These markings are very important for your safety. Each electrical device in your home, from the main service panel to the cover plate on the light switch, must bear one of these markings to show it has been properly tested and is certified.   It may be tempting to pick up that cheap, uncertified, light fixture in Mexico or elsewhere, or that under the counter electrical gizmo at a fraction of cost for the same thing back home but don't do it. It may be a fire hazard or a shock hazard. Manufacturing to a safe standard and submitting that equipment to a laboratory for rigid tests is an expensive process but it is your assurance that that device meets certain minimum standards. Always look for one of the above certification marks on all electrical equipment you purchase. If you do not find any of these markings bring the device to the attention of the store manager.

3   ELECTRICAL INSPECTIONS   WHEN TO CALL FOR INSPECTION.

ROUGH WIRING INSPECTION - Ontario Bulletin 12-1-6

Before calling for the rough wiring inspection make sure that:

1.   **The electrical service** equipment is in place. The service conduit or cable, the meter base, service panel and the service grounding cables should all be installed for the rough wiring inspection.

   Note   If you are using service conduit, not cable, the conductors need not be installed but the conduit must be in place.

2.   **All branch circuit cables** are in place and properly strapped and protected from driven nails. Don't forget, all joints, splices and ground wire connection to boxes must be completed as far as possible.

Avoid Costly Rejections and Delays.

   Before calling for the rough wiring inspection check your work very carefully as an Inspector would do his inspection. Check every detail. If you have followed the instructions in this book you should do well.

Caution - Do not cover any wiring, not even with insulation, until it has been inspected and approved for covering, Rule 2-000(7).

   Note -   In some rural areas a sketch map may be necessary to help the Inspector find your premises. Also, if a key is required, provide instructions where to find it.

FINAL INSPECTION - Ontario Bulletin 12-1-6

I.   **Your installation must be complete.** It's a good idea to check it carefully as an Inspector would do his inspection. Use this book as you go through your installation to check off each item. Look for forgotten or unfinished parts. Check for such things as circuit breaker ratings, tie-bars, circuit directory, grounding connections etc.. Look for open KO holes and unfinished outlets. Don't forget, if any of the appliances such as a dish washer is not going in just yet you must terminate the supply cable in a fixed j-box, (the box must be fastened in place) with cover until that appliance is actually installed.

2   **Electrical Permit - Check also if your permit covers all you have installed.**

   The Inspector is one of the good guys. His concern is that your installation is safe and that it meets minimum code requirements. That is, after all, exactly what you want too.

4   ELECTRICAL INSPECTION OFFICES

Mississauga .... Tel. 905 712 5650, 1 800 434 0172
                        Fax 905 507 8204,   1 800 434 0173
                        155A Matheson Blvd. W.
                        Mississauga Ont. L5R 3L5

North. Metro .... Tel. 905 946 6297, 1 800 434 0174
                        Fax 1 800 434 0175
                        185 Clegg Rd.
                        Markham, Ont. L6G 1B7

Toronto .... Tel. 905 946 6297, 1 800 434 0174
                        Fax 1 800 434 0175
                        10 Gateway Bld., Suite 410.
                        Don Mills, Ont. M3C 3A1

Bancroft .... Tel 613 969 0201 & 1 800 369 7536
                        Fax 613 968 3085 & 1 800 369 7539
                        Hwy #28 South, Box 310, K0K 1C6

Belleville .... Tel 613 969 0201, Fax 613 968 3085
                        Tel 1 800 369 7536,   Fax 1 800 369 7539
                        100 Bell Blvd., Suite 330
                        Belleville, Ont. K8P 4Y7

Cornwall .... Tel 613 225 7600, 1 800 369 7535
                        Fax 613 225 9101,   1 800 369 7542
                        Second Street West
                        Box 999, Cornwall, Ont. K6H 5V1

Kingston .... Tel 613 969 0201, 1 800 369 7536
                        Fax 613 968 3085, 1 800 369 7539
                        211 Counter Street, F1
                        Kingston, Ont. K7L 4X7

Ottawa .... Tel 613 225 7600, 1 800 369 7535
                        Fax 613 225 9101,   1 800 369 7542
                        106 Colonnade Rd. N. Suite #210
                        Nepean, Ont K2E 7L6

Renfrew .... Tel 613 225 7600, 1 800 369 7535
                        Fax 613 225 9101,   1 800 369 7542
                        249 Raglan St. S. , 2nd Fl. Post Office Bldg.
                        Renfrew, Ont. K7V 1R3

Smith Falls .... Tel 613 225 7600, 1 800 369 7535
                        Fax 613 225 9101,   1 800 369 7542
                        R.R.#4, Hutcheson Rd.
                        Smith Falls, Ont. K7A 4S5

Kenora .... Tel 705 495 4541, 1 800 636 7107
                        Fax 705 494 4331,   1 800 636 7108,
                        6 - 621 Lakeview Drive .
                        Kenora, Ont. P9N 3P6

Manitoulin .... Tel 705 495 4541, 1 800 636 7107
                        Fax 705 494 4331, 1 800 636 7108
                        Service Centre, Hwy 6
                        Little Current, Ont. P0P 1K0

New Liskeard .... Tel 705 495 4541, 1 800 636 7107
                        Fax 705 494 4331, 1 800 636 7108
                        221 Sheperdson Rd., Box 490
                        New Liskeard, Ont. P0J 1P0

North Bay .... Tel 705 497 7378, 1 800 636 7107
                        Fax 705 494 4331, 1 800 636 7108
                        1275 Main St. West
                        North Bay, Ont. P1B 2W7

Sault Ste Marie ..... Tel 705 495 4541, 1 800 636 7107
                        Fax 705 494 4331, 1 800 636 7108
                        71 Black Rd. Unit 7
                        Sault Ste Marie, Ont. P6A 6J8

Sudbury .... Tel 705 495 4541, 1 800 636 7107
                        Fax 705 494 4331,   1 800 636 7108
                        1760 Regent St. South
                        Sudbury, Ont. P3E 3Z8

Thunder Bay .... Tel 705 495 4541, 1 800 636 7107
                        Fax 705 494 4331, 1 800 636 7108
                        34 North Cumberland St., 4th. Floor
                        Thunder Bay, Ont. P7A 4L5

Timmons .... Tel 705 495 4541, 1 800 636 7107
                        Fax 705 495 4331 1 800 636 7108
                        Mountjoy Street South, Box 340
                        Timmons, Ont. P4N 7C8

Brantford .... Tel 905 529 7125, 1 800 278 4264
                        Fax 905 529 1890, 1 800 545 8879
                        97 Brant Avenue
                        Brantford, Ont., N3T 3H4

Hamilton................. Tel 905 529 7125, 1 800 278 4264
                        Fax 905 529 1890,   1 800 545 8879
                        989 King Street West
                        Hamilton, Ont., L8S 1B9

Kitchener .... Tel 519 746 3040, 1 800 813 5482
                        Fax 519 746 3041, 1 800 667 4278
                        279 Weber Street, North
                        Waterloo, Ont.., N2J 3H8

London .... Tel 519 680 4565, 1 800 813 5663
                        Fax 519 680 4542, 1 800 223 2029
                        1071 Wellington Rd.. South
                        London, Ont. N6E 1W4

Stratford .... Tel 519 746 3040, 1 800 813 5482
                        Fax 519 746 3041, 1 800 667 4278
                        40 Boyd Street
                        Stratford, Ont. N5A 6S4

Thorold .... Tel 905 227 5380, 1 800 813 8467
                        Fax 905 227 2711,   1 800 545 9572
                        40 Albert Street, W.
                        Thorold, Ont. L2V 2G3

Wallaceburg .... Tel 519 680 4565, 1 800 813 5663
                        Fax 519 680 4542,   1 800 223 2029
                        30 McNaughton Avenue, Unit E-1
                        Wallaceburg, Ont. N8A 1R8

Windsor .... Tel 519 972 3110, 1 800 880 9463
                        Fax 519 972 1888,   1 800 223 2034
                        3357 Walker Rd.
                        Windsor, Ont. N8W 5J7

Barrie .... Tel 705 726 5169, 1 800 571 7724
                        Fax 705 726 2862, 1 800 571 7725
                        48 Alliance Blvd., Box 4300
                        Barrie, Ont. L4M 6L4

Bracebridge .... Tel 705 726 5169, 1 800 571 7724
                        Fax 705 726 2862, 1 800 571 7725
                        Taylor Court, Box 810
                        Bracebridge, Ont. P1L 1V1

Durham .... Tel 705 745 3236, 1 800 305 7383
                        Fax 705 745 2046, 1 800 749 2016
                        420 Green Street, Unit #5
                        Whitby, Ont. L1N 8R1

Fenlon Falls ....Tel 705 745 3236, 1 800 305 7383
                        Fax 705 745 2046, 1 800 749 2016
                        Box 550, Glenarm Rd.
                        Fenlon Falls, Ont. K0M 1N0

Newmarket .... Tel 705 745 3236, 1 800 305 7383
                        Fax 705 745 2046,   1 800 749 2016
                        1124 Stellar Drive
                        Newmarket, Ont. L3Y 7B7

Orangeville .... Tel 705 726 5169, 1 800 571 7724
                        Fax 705 726 2862,   1 800 571 7725
                        125 "C" Line
                        Orangeville, Ont. L9W 3V2

Owen Sound .... Tel 705 726 5169, 1 800 571 7724
                        Fax 705 726 2862, 1 800 571 7725
                        1025 - 6th. Street E.
                        Owen Sound, Ont.., N4K 5P6

Peterborough .... Tel 705 745 3236, 1 800 305 7383
                        Fax 705 745 2046, 1 800 749 2016
                        749 The Kingsway
                        Peterborough, Ont., K9J 6W7

Walkerton .... Tel 705 726 5169, 1 800 571 7724
                        Fax 705 726 2862, 1 800 571 7725
                        Bruce County Road 3, Box 1390
                        Walkerton, Ont., N0G 2V 0

5   SERVICE SIZE - Rule 8-200

We must begin with the electrical service box. It may appear to be very difficult to install a new electrical service in your house but it need not be so. If you read these instructions carefully it should be easy and enjoyable and what's more, you should save a bundle.

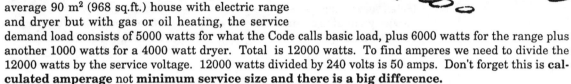

The rules permit a service as large as we like but not as small as we like. There is a definite minimum size we must have if it is going to be passed by the Inspector.

Service size is based on two things.

1.   **Calculated Load** - This is the sum of all the loads after certain demand factors are applied. For an average 90 m² (968 sq.ft.) house with electric range and dryer but with gas or oil heating, the service demand load consists of 5000 watts for what the Code calls basic load, plus 6000 watts for the range plus another 1000 watts for a 4000 watt dryer. Total is 12000 watts. To find amperes we need to divide the 12000 watts by the service voltage. 12000 watts divided by 240 volts is 50 amps. Don't forget this is **calculated amperage** not **minimum service size and there is a big difference.**

2.   **Minimum Service Size** — This is based on floor area.

> **60 amp.** —   For any house with **LESS THAN** 80 m² (861 sq.ft.) floor area. This includes the areas of all the floors except the basement. Basement floor area is ignored completely for this purpose.

> **100 amp.** —   For any house with 80 m² (861 sq. ft.) **OR MORE** floor area. As above, this includes all floors except basement.

>> **Note** — Even if the actual load is, say 30 amps, the minimum size service permitted is 100 amps if the floor area is 80 m² or more. This extra capacity is for future load.

>> The following table has been developed to simplify the service size calculation. The table shows a progressive load - in other words, any service size given in any column assumes all the loads shown above that value are going to be used.

>> For example, a 90 m² (no basement) house with no electrical appliances such as a range, dryer, water heater or electric heating would require a 100 amp service. The material required for this service is listed under B, page 6.

>> If the 90 m² no basement house had a 12,000 watt range, 4000 watt dryer, 3800 watt water heater, the service size would still be only 100 amps. This is the minimum size service for this floor area. The material required for this service is listed under B, page 6.

>> If the 90 m² no basement house had a 12000 watt range, a 4000 dryer, a 3800 watt water heater and 15 kw. electric hot water boiler the service size would need to be minimum 117.1 amps. The material required for this service is listed under C, page 6.

>> Once you have determined the correct list of material for your house, measure the lengths required for your job. You can use this list when shopping for the materials you will require.

LIST of MATERIALS

Remember you need twice as much hot conductor length as neutral conductor.

A   **Service Size - 60 amps.**   Note - may be used only if floor area above ground is less than 80 m² (861 sq. ft.).

        Service switch, fuse or breaker rating ---------- 60 amps.
        Hot conductors-------------------------------------- 2 - #6 R90XLPE copper (black, red or blue)
        Neutral conductor --------------------------------- 1 - #6R90XLPE copper (white)
        Service conduit ------------------------------------- 1 inch.
          or use #6 copper TECK cable.
        Meter base rating ---------------------------------- 100 amps.
        Service grounding conductor --------------------- #6 (or larger) bare copper.
        Service panel size --------------------------------- 16 circuits (minimum)

B   **Service Size - 100 amps.**

        Service switch, fuse or breaker rating ---------- 100 amps.
        Hot conductors ------------------------------------- 2 - #3 R90XLPE copper (black, red or blue)
        Neutral conductor ---------------------------------- 1 - #3 R90XLPE copper (white)
        Service conduit ------------------------------------- 1¼ inch.
          or use #3 copper TECK cable.
        Meter base rating ---------------------------------- 100 amps.
        Service grounding conductor --------------------- #6 (or larger) bare copper.
        Service panel size --------------------------------- 24 circuits.

    Note   This panel may supply a central electric furnace, an electric boiler or baseboard heaters.

C   **Service Size - 120 Amps.**

        Service switch, fuse or breaker rating ---------- 125 amps.
        Hot conductors ------------------------------------- 2 - #2 R90XLPE copper (black, red or blue)
        Neutral conductor ---------------------------------- 1 - #3 R90XLPE copper (white)
        Service conduit ------------------------------------- 1 ¼ inch.
          or use #2 copper TECK cable.
        Meter base rating------------------------------------200 amp.
        Service grounding conductor ---------------------- #6 (or larger) bare copper
        Service panel size --------------------------------- 24 circuits

    Note   This panel may supply a central electric furnace or electric boiler. If heating is with electric baseboards you will need more than 24 branch circuits, see list D below.

D   **Same as C above except:**

Service Panel Size - 30 circuits.

    Note   This panel may supply a central electric furnace, an electric boiler or electric baseboard heaters.

E   **Service Size - 150 Amps.**

        Service switch, fuse or breaker rating . 150 amps.
        Hot conductors .... 2 - 31/0 R90XLPE copper (black, red or blue)
        Neutral conductor .... 1- #3 R90XLPE copper (white)
        Service conduit .... 1¼ inch.
          or use #110 copper Teck cable.
        Meter base rating .... 200 amp.
        Service grounding conductor................ #4 (or larger) bare copper
        Service panel size................................. 30 circuits

Note   This panel may supply a central electric furnace or electric boiler. If heating is with electric baseboards you will need more than 30 branch circuits, see list F below.

F   **Same as E above except:**

Service Panel Size ----------------------------- 40 circuits.

Note   This panel may supply central electrical furnace, an electric boiler or baseboard heaters.

G   **Service Size - 200 Amps.**

Service switch, fuse or breaker rating . 200 amps.
Hot conductors .... 2 - #2/0 R90XLPE copper (black, red or blue)
Neutral conductor .... 1 - #3 R90XLPE copper (white)
Service conduit .... 1¹/₂ inch.
or use #310 copper Teck cable.
Meter base rating .... 200 amps.
Service grounding conductor .... #3 (or larger) bare copper.
Service panel size .... 30 circuits.

Note   This panel may supply a central electric furnace or electric boiler. If heating is with electric baseboards the 30 circuit panelboard is too small. You will need to install a 40 circuit panelboard.

NOTES   NOTES   NOTES

A   **Bare Neutral** - Rules 4-020, & 6-306 - All the lists of material above refer to an **insulated neutral**. The rules do not require the neutral to be insulated, it may be bare. Using a bare neutral will save a bit of money and it is easier to form into the desired shape to make connections but the Code does not allow us to reduce our service conduit size even though it is smaller than an insulated conductor. Rule 12-1014(4)(e) says bare conductors require the same conduit size as insulated conductors.

Bulletin 4-5-4 says that where the bare neutral enters a meter base or switch or a panel it must be insulated to prevent contact with live parts. In most cases it should be possible to train this bare conductor so that there is no danger of inadvertent contact with live parts. However, where it is not possible to prevent inadvertent contact with live parts the bare conductor must be insulated. Apply a layer of electrical tape throughout the length of the exposed section. This layer of tape should be equal to the thickness of the insulation on the hot conductors.

B   **Spare Circuits - 2 Required** - Rule 8-108(2) requires that there be at least two spare circuits left in the panel after you have connected all the circuits you have installed. These two circuits are for future use. See also under (b) "How Many Circuits Do I Need", on page 34.

C   **Minimum Panel Size** - The above lists give the minimum panel size required by code in each case, however, this may not be enough for your installation. Make sure that you have a sufficient number of circuit spaces available in your panel. See also under (b) "How Many Circuits Do I Need", on page 34.

D   **Caution - Minimum Service Conductor Size** - Bulletin 4-5-4 - The above lists of material are based on a literal interpretation of Rule 4-022, which permits a reduced size service neutral conductor. While there is general agreement in the electrical trade, that a full size neutral conductor is wasteful because it cannot possibly ever be used, there is disagreement on how much reduction should be permitted. For this reason you should consult your local Inspector if he will accept a reduced neutral conductor.

The following explanation will help the student understand the basis of the above lists of material.

Consider the following:

It is argued that Rule 8-200 specifies a minimum size for service conductors for residences and the point is stressed that this includes the neutral therefore, as the argument goes, we may not reduce the neutral below the minimum size stated in that rule. It should be noted that almost the same words are used in Rule 8-202, 8-204 & 8-208. Each of these Rules deals with different buildings and each states "the minimum ampacity of service conductors" may not be smaller than permitted in that rule. In each case the "service conductors" includes the neutral conductor, therefore, as the argument goes, we may not apply Rule 4-022 in any of those cases either. The effect of this interpretation is that we may never reduce the neutral conductor for any service. In fact, most of these rules refer to both the "service" and the "feeder" conductors. Because the neutral is a "feeder conductor" this interpretation of the rule would, therefore also not permit it's size to be reduced under any conditions.

It should be noted that one of the basic rules of interpretation is that a specific rule supercedes or overrides a general rule. Rule 4-022 is such a specific rule. Rule 8-0200 is specific in that it refers to "service conductors" as opposed to "branch circuit conductors" but Rule 4022 is more specific in that it refers to the neutral only. Then notice, it refers to the neutral conductor in the service. It should also be noted that this rule does not limit the application to commercial buildings nor does it limit in any way to services above 100 amperes only. Rule 4-022 says that we do not need a full size neutral in a residential service. The rule says it need only be large enough to carry all the load connected to the neutral. It goes on to say that this load must be determined by the minimum demands required by Rule 8-200.

In a residence the loads connected to the neutral consist of lights, plugs, and in some cases a few range cooking top elements. The oven, electric heating, dryer, water heater and sauna loads are all not connected to the neutral. Applying all the demand factors permitted by the rules the calculated load on the neutral for an 969 sq. ft. (90 m^2) house would be 5000 watts. This is 20.8 amperes. Therefore, according to code. the connected load on the neutral in this house is 20.8 amperes. The smallest size neutral specified in the above lists is #6. A #6 - R90 copper conductor may carry 65 amperes. This leave 65 - 20.8 = 44.2 amperes spare neutral capacity. Even if we added a 12kw range, and it would be incorrect to do so because it is not connected to the neutral, the service neutral load would still be less than the #6 copper service neutral could carry. The important thing to remember is that the service neutral must be the same size as the service grounding conductor. Don't waste your money on a full size neutral and conduit when it is simply not possible to ever use it in a residence. Spend your money on things that can be used to improve your installation. More plugs on the kitchen and the basement areas are a much better investment.

Service Size Table - For HOME OWNER USE

Connected Load	If total, of all floor areas except basement, is less than 80 m² (861 sq. ft.)		If total, of all floor areas except basement, is between 80 m² and 90 m² (969 sq.ft.)		If floor area is between 90 m² and 180 m² (1937 sq.ft.)	If floor area is between 180 m² and 270 m² (2906 sq.ft.)	If floor area is between 270 m² and 360 m²(3875 sq.ft.)
	No basement house	Why any size basement	No basement house	With any size basement	See notes 2 & 5	See Note 2	See note 2
Basic Load only See Note 3	60 Amps A	60 Amps A	100 Amps B	100 Amps B	100 Amps B	100 Amps B	100 Amps B
Plus range (up to 12 KW rating)	60 Amps A 60 Amps A	60 Amps A 60 Amps A	100 Amps B 100 Amps B	100 Amps B 100 Amps B	100 Amps B 100 Amps B	100 Amps B 100 Amps B	100 Amps B 100 Amps B
Plus 4 KW dryer	60 Amps A	60 Amps A	100 Amps B	100 Amps B	100 Amps B	100 Amps B	100 Amps B
Plus 3 KW water heater See Note 8.	60 Amps A	60 Amps A	100 Amps B	100 Amps B	100 Amps B	100 Amps B	100 Amps B
Plus heating with electric hot air furnace or electric hot water boiler							
10kw 15 kw	96.1 amps B 117.1 amps C	100.3 amps B *121.3 amps E	96.1 amps B 117.1 amps C	100.3 amps B *121.3 amps E	100.3 amps B *121.3 amps E	*104.5 amps C 125.5 amps E	108.6 amps C 129.6 amps E
18 kw 20 kw	130.1 amps E 138.1 amps E	134.3 amps E 142.3 amps E	130.1 amps E 138.1 amps E	134.3 amps E 142.3 amps E	134.3 amps E 142.3 amps E	138.5 amps E 146.5 amps E	142.6 amps E *150.6 amps E
24 kw 27 kw			*156.1 amps G 169.1 amps G	160.3 amps G 173.3 amps G	160.3 amps G 173.3 amps G	164.5 amps G 177.5 amps G	168.6 amps G 181.6 amps G
30 kw			185.3 amps G	185.3 amps G	185.3 amps G	189.5 amps G	193.6 amps G
Or if using Baseboard heaters Sum of all heater ratings							
4 kw 5 kw 6kw	69.8 amps B 74.0 amps B 78.1 amps B	100.0 amps B 100.0 amps B 100.0 amps B	100.0 amps B 100.0 amps B 100.0 amps B	100.0 amps B 100.0 amps B 100.0 amps B	100.0 amps B 100.0 amps B 100.0 amps B	100.0 amps B 100.0 amps B 100.0 amps B	100.0 amps B 100.0 amps B 100.0 amps B
7 kw 8 kw 9 kw	82.3 amps B 86.5 amps B 90.6 amps B	100.0 amps B 100.0 amps B 100.0 amps B	100.0 amps B 100.0 amps B 100.0 amps B	100.0 amps B 100.0 amps B 100.0 amps B	100.0 amps B 100.0 amps B 100.0 amps B	100.0 amps B 100.0 amps B *103.1 amps D	100.0 amps B 100.0 amps B *103.1 amps D
10 kw 11 kw 12 kw	94.8 amps B 97.9 amps B *101.0 amps D	100.0 amps B *102.1 amps D 105.2 amps D	100.0 amps B 100.0 amps B *101.0 amps D	100.0 amps B *102.1 amps D 105.2 amps D	100.0 amps B *102.1 amps D 105.2 amps D	*103.1 amps D 106.3 amps D 109.4 amps D	107.3 amps D 110.4 amps D 113.5 amps D
13 kw 14 kw 15 kw	*104.2 amps D 107.3 amps D 110.4 amps D	108.3 amps D 111.5 amps D 114.6 amps D	*104.2 amps D 107.3 amps D 110.4 amps D	108.3 amps D 111.5 amps D 114.6 amps D	108.3 amps D 111.5 amps D 114.6 amps D	112.5 amps D 115.6 amps D 118.8 amps D	116.7 amps D 119.8 amps D *122.9 amps E
16 kw 17 kw 18 kw	113.5 amps D 116.7 amps D 119.8 amps D	117.7 amps D 120.8 amps E 124.0 amps E	113.5 amps D 116.7 amps D 119.9 amps D	117.7 amps D 120.8 amps E 124.0 amps E	117.7 amps D *120.8 amps E *124.0 amps E	*121.9 amps E *125.0 amps E 128.1 amps E	126.0 amps F 129.2 amps F 132.3 amps F
19 kw 20 kw	*122.9 amps E 126.0 amps F	127.1 amps F 130.2 amps F	*122.9 amps E 126.0 amps F	127.1 amps F 130.2 amps F	127.1 amps F 130.2 amps F	131.3 amps F 134.4 amps F	135.4 amps F 138.5 amps F

The table gives the minimum service ampacity required in each case. The letter after each ampere rating indicates which list of materials to use. For Sauna Heaters see Note 10, page 10.

These notes refer to references on the Table.

(1)   **Remember,** this table gives the **minimum service sizes permitted** under the rules. Services of higher rating may be installed and sometimes may be an advantage for future load additions.

(2)   **The floor areas** in all these columns include the floor area of an in-house garage but does not include an open carport. Calculate your floor area by adding basement floor area at 75% and the other floor areas at 100%.

(3)   **Basic load** - This includes:
- All lighting outlets
- All 15 amp plug outlets
- Hot air furnace (standard) gas or oil burning type.
- Any appliance of less than 1500 watts (this is the same as 12.5 amps at 120 volts) each. This includes loads such as a garburator, freezer, toaster, etc. but does not include **fixed** (permanently wired not plug in type) electric heating.

(4)   *** Some amperages shown** on the table are marked with an asterisk. This means that the next size smaller service **may** be acceptable in this case **by special permission.** Please note this special permission is not automatic - you must request it from your local Inspector and he may refuse it if he thinks it too small, Rule 8-106(1). In any case, it is not likely you will be permitted to use a smaller branch circuit panel.

(5)   **For example,** if the floor area is 90.1 m^2 we must use this column.

(6)   **The table may not show** the exact rating of your electrical load - in that case you may take the next size larger size or calculate in detail, as described on the next page.

(7)   **Motor loads** - Hot air and hot water heating systems require motors to circulate the heat. These motors are small, they require approx. 7 amps at 120 volts. This is included in the values given in the table.

(8)   **6 kw Water Heater** - This does not include an electric water heater for a hot tub or spa. Most water heater tanks are equipped with 2 - 3 kw heater elements. While this is 6000 watts in total, the switching arrangement in the thermostat is such that only one 3 kw element is working at any one time. It is a flip-flop switching arrangement. Under normal water use the lower 3 kw element will heat the water. When the demand for hot water becomes too great for the lower element the thermostat disconnects the lower heater element and connects only the upper heater element. The upper element will heat the water in only the top part of the tank, this provides rapid recovery of hot water. When the demand for hot water decreases the thermostat will switch to the lower heater element again.

(9)   **7000 Watt Water Heaters** - There is an energy saving tank available. It has 2 - 3800 watt heater elements, total wattage is 7600. With the flip-flop thermostat, as described above, the maximum load on this tank is 3800 watts. This is 800 watts more than allowed for in the table. For service calculations we add only 25% of this 800 watts. The additional load is only 200 watts. In an exam this is important - even this little bit - but on the job - - - well, it's not even one ampere more.

    Swimming pool, Hot tub or spa - Electric water heaters for these loads are not included in the table. These must be added at 100% of their rating, Rule 8-200(1)(v).

(10)   **Sauna** - Rule 62-102 says this must be added as fixed electric space heating.

    If you are using gas or oil (not electricity) to heat your house, enter the table with the full rating of this heater as if it were an electric baseboard heater. For example, if you are installing a 4.5 kw sauna, you would enter the table as if it was a 5kw baseboard heater then read across under the correct floor area. Baseboard heaters are listed in the lower left corner of the table.

    If you are using electric baseboard heaters, then simply add the sauna kw load to the total baseboard heater load before you enter the table.

    If your are using an electric hot air furnace use the table for all the other loads, then add the sauna load. All the values given in the table are in amperes, therefore, we need to convert the sauna load to amps as well. Do this by dividing the sauna watts by 240 volts then add this to the other load amps. Total amps is the minimum service size. Then use the list of materials on page 6.

(11)   **An electric range plug** outlet must be installed in every house. This is not a choice, it's required by Rule 26-746(5).

(12)   **Built-in Vacuum Cleaning System** - Most of these systems will draw 12.5 amps or less at 120 volts. In that case they are included in the basic load shown in the table. Those units which draw more than 12.5 amps at 120 volts must be added at 25% of their rating but remember to take only half of the amperage because the service is calculated at 240 volts. For example, if your service should be 125 amp according to the above table and you want to add a 14.0 amp vacuum cleaner system, it would look like this:

```
Other load ------------------------------------------------------------- 125.00 amps.
Vacuum system ---- 25% of 14 divided by 2 = 0.25 X 7 =     1.75 amps
                                               Total      =   126.75 amps
```

    Hardly worth the effort but it could mean that the next size larger service conductors may need to be used.

Now we know how large the service must be.

Next step, refer to Lists of Material on page 6.

DETAILED CALCULATION FORM   (for student use)

Use the following format. Fill in the blanks as required to describe your installation. This calculation gives minimum size service required. Then refer to the appropriate Lists of material on page 6.

Step 1   **Basic Load**   Rule 8-200(1)

1st. 90 m² floor area.. =   5000 watts
Next 90 m² floor area or portion thereof .... (Add 1000 watts) .... =_____ watts
Next 90 m² floor area or portion thereof .... (Add 1000 watts) .... =_____ watts

Note (1) Floor area in this case must include 75% of the basement floor area plus 100% of all other living floor areas. All floor areas are inside, actual, floor area measurements.

Note (2) This basic load includes all lighting and plug outlet loads. It includes oil or gas furnace and any other appliances such as built-in vacuum systems (which are rated 12.5 amps or less), swimming pool pump motors, most workshop motors, compactor motors, garburators, air conditioners, each individually rated at not more than 1500 watts (this is 12.5 amps at 120 volts) but does not include fixed electric space heating.

Step 2   **Appliances**

Range (For a 12 kw or smaller range) - add 6000 watts .... =_____ watts
   Plus - (If it is greater than 12kw, add) 40% of that part which is in excess of 12kw .... =_____ watts
      Note - 6000 watts is not a percentage of the range rating - it would be 6000 watts for any size range up to 12kw.

2nd. Range - Add 25% of its wattage. (25% of 12000 for a 12kw range) .... =_____ watts
      (See App. to Rule 8-200, Page 577 in Code.)
Dryer - Add 25% of its rating if a range is provided for .... =_____ watts
Water heater - Add 25% of rating if a range is provided for .... =_____ watts

Note - If this is all the load we have, i.e. if heating is with gas or oil and there is no other large load such as electric sauna etc. Then we must determine minimum service size here as follows:

(a) **If the floor area is less than 80 m²**, this includes the area of all the floors but does not include the basement. Basement floor area is ignored completely for this determination. **Then** - minimum service size must be 60 amps. Now see List of Material, page 6.

(b) **If the floor area is 80 m²** or more, for this determination, as in (a) above, we may ignore completely the basement floor area. **Then** - minimum service size must be 100 amp. Now see List of Material, page 6.

Step 3   **Sauna** - (The Code calls this space heating).

Add sauna load at 100% to other heating loads if house heating is not with electricity............. =_____ watts
If the house is electrically heated see under specific type of heating below.

Step 4   **Electric Baseboard Heating & Sauna**

Baseboard heaters - add watts of all heaters .... =_____ watts
Sauna heater .... =_____ watts
                                                                    _____
                            Total =   _____ watts

1st. 10 kw must be added at 100% .... =_____ watts
All the balance may be added at 75% .... =_____ watts
                                                                                                                    _____
                                                            Total =_____ watts

Thus far all our calculations are based on watts because that is what many of the demand factors in the Code are based on. The balance of our calculations, however, are based on amperes not watts. At this point it is best to convert the total watts we have gotten so far into amperes, then we can complete the balance of our calculation using only amperes.

$$\frac{\text{Total watts}}{240 \text{ volts}} = \text{...} \underline{\hspace{1cm}} \text{ amps}$$

The following loads are usually added directly in amperes.

Step 5   Electric Hot Air Furnace & Sauna

Add 100% of furnace nameplate rating .... = \underline{\hspace{1cm}} amps

Add sauna at 75% of its nameplate rating .... = \underline{\hspace{1cm}} amps

Step 6   Air Conditioning .... = \underline{\hspace{1cm}} amps

Total amps .... = \underline{\hspace{1cm}} amps .... \underline{\hspace{1cm}} amps

Minimum service size is \underline{\hspace{1cm}} amps

Now consult the List of Materials on page 6.

EXAMPLE 1

An average 120 m² (1291sq. ft.) house. It has a 60 m² (646 sq. ft.) basement. The electrical load consists of 14kw. range upstairs and a 12 kw range in the basement, 4 kw. dryer, 3 kw. water heater, 3 kw. sauna and 11kw. electric baseboard heating. There is also a 12 amp A/C unit. Calculate service size.

Basic Load - Floor area = 120 m² main floor at 100% =   120 m²

60 m² basement at 75% =   \underline{45 m²}

Total =   165 m²

First 90 m²	5000 watts
Next 75 m²	1000 watts
Range	6000 watts
Plus 40% of 14kw -12kw = ...	800 watts
2nd range at 25% of rating	3000 watts
Dryer	1000 watts
Water heater	750 watts

Electric heating & sauna = 11 kw. plus 3 kw. = 14 kw.

First 10 kw. at 100% .... 10000 watts

Balance, 4 kw. at 75% .... 3000 watts

Total =   30550 watts

$$\frac{30550}{240} = \text{...} 127.3 \text{ amps}$$

12 amp A/C unit Add this at 100% rating... 12.0 amps

Minimum service size required   139.3 amps

Now refer to the lists of material on page 6. For this service use list F.

Our load is 139 amps. This is greater than list C, which is 120 amps, but less than E, which is 150 amps. Therefore we may use list E

CAUTION - Don't ignore the note regarding the panel. The panel for list E may not supply baseboard heaters - but that is what this question requires. Therefore, we must use list F because we need a 40 circuit panel for this load.

SUB-FEEDER SIZES   to 2nd. PANEL

It is often an advantage to install a second panel near the kitchen load. The size of panels and feeders and other details are dealt with on page 36.

6 SERVICE CONDUCTOR TYPES - Rules 12-100 & 12-102

(a) Copper / Aluminum Conductors

The service conductor may be either copper or aluminum. Aluminum conductor installation requires great care and is not covered in this book.

(b) Service Conductor Insulation

Service conductors are subjected to extreme temperature changes. For this reason the insulation on these conductors must be approved and marked to show it is certified for use in low temperature locations. Look for surface printed marking on the conductors. It will say "Minus 40°C" or "-40°C". You may use conductors such as TW-40 or R90 or RW 90 or others which have the -40°C marking.

The reason for this restriction is that service conductors, if connected to overhead lines are required to move, even if only slightly. During cold winter temperatures this small flexing may cause hairline cracks in the conductor insulation. Such conductor insulation becomes useless as a protection against electrical shock. Accidental contact with such a conductor when it is wet could result in severe electrical shock.

7   HYDRO SERVICE WIRES to the HOUSE

(a)   **Consult Hydro** - Rules 6-112, 6-116, 6-206 & Ontario Bulletin 6-1-4

Before any work is done the power utility should be consulted to determine which pole the service will be from. This is very important. The entrance cap must be properly located with respect to the Hydro pole.

There are a number of details to watch out for when locating the service entrance cap:

A   **Entrance Cap above Line Insulator** - Rule 6-116(b) requires the entrance cap to be located between 6 in. and 12 inches above the line insulator. This is to prevent water from migrating along the service conductors to the meter base. See also page 16 for more details on this requirement.

B   **Roof Crossing** - Rule 12-312 says that conductors shall not cross over any buildings except by special permission from your local Electrical Inspector. Locate the service entrance cap so that Hydro service leads may run directly to it without having to cross over any roof except the roof overhang as shown in the illustration below.

C   **Heavy Snowfall Areas** - The three locations shown along the side of the roof in the illustration below may not be acceptable in heavy snowfall areas. Check with your local Inspector.

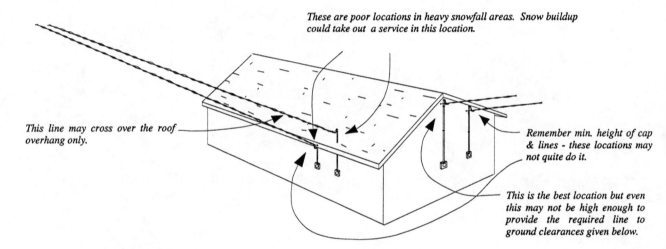

These are poor locations in heavy snowfall areas. Snow buildup could take out a service in this location.

This line may cross over the roof overhang only.

Remember min. height of cap & lines - these locations may not quite do it.

This is the best location but even this may not be high enough to provide the required line to ground clearances given below.

The illustration above shows some service locations which are acceptable and some which are not acceptable,

(b)   **Minimum Line to Ground Clearances** - Rule 6-112(2)

The insulator, shown below, required by Hydro for their service drop, must be installed high enough to provide the following minimum clearances:

	Meters	Feet
Any Public roadway	6.0	19.7
Any Private Roadway or Driveway and any other space, including lawns, which are accessible to:		
Commercial & Farm Vehicles	6.0	19.7
Pedestrians & Passenger Vehicles only	4.5	14.8
Sundeck	2.5	8.2

(c)   **Sundeck Crossing** - Rule 12-310 & 12-312

The strict interpretation of the rules requires special permission for service leads to cross over a sun deck. Such permission would not likely be granted unless the lines are at least 2.5 m (8.2 Ft.) above the deck.

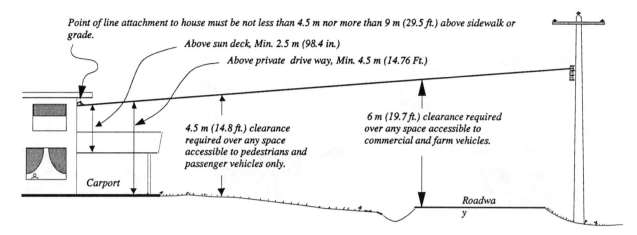

Point of line attachment to house must be not less than 4.5 m nor more than 9 m (29.5 ft.) above sidewalk or grade.

Above sun deck, Min. 2.5 m (98.4 in.)

Above private drive way, Min. 4.5 m (14.76 Ft.)

4.5 m (14.8 ft.) clearance required over any space accessible to pedestrians and passenger vehicles only.

6 m (19.7 ft.) clearance required over any space accessible to commercial and farm vehicles.

Carport

Roadway

(d)   **Hydro Line Attachment to House** - Rule 2-108, 6-112 & 6-116

Hydro requires that you provide and install an insulator on the building for their line crew to attach their service cable. This wireholder must:

— Be insulated type, even for triplex cable.

— Be located within 24 in. of the entrance cap.

— Be high enough to maintain all the line to ground clearances given above but must not be located higher than 29.5 ft above the sidewalk or grade level..

— Be carefully located so that the 1 m (39.4 in.) line to window and door clearances can be maintained. See illustration on Page 17.

— Be sufficiently well anchored in a structural member of the building to withstand the pull of the lines in a storm.

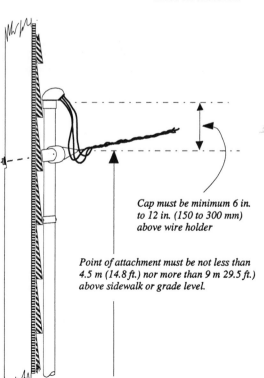

Cap must be minimum 6 in. to 12 in. (150 to 300 mm) above wire holder

Point of attachment must be not less than 4.5 m (14.8 ft.) nor more than 9 m 29.5 ft.) above sidewalk or grade level.

Note   Rule 6-112(6) requires that the wire holder must be bolt type as shown above.

This means , of course, that if a tree falls across the line to your house or if an overheight 70 to truck comes hurtling down the road and hooks onto the Hydro lines to your house you are safe - your wire holder is designed to withstand these kinds of stress. . . . . . It is possible the stress will pull out the wall but you can count on that bolt to not let go of the 2 x 4 stud. (It's best to choose a wall that could be pulled out without it taking down the whole house with it.)

Note -   The construction of the wire holder. It must be a type which is designed to hold the lines so they cannot fall to the ground even if the porcelain insulation material on the insulator should break.

8   SERVICE ENTRANCE CAP - For overhead services only.

(a)   **Consult Hydro** - Rules 6-112, 6-116(a) & 6-206 & Ontario Build Code 9.35.5

Before beginning the installation of your electrical service be sure to obtain from the supply authority (Ontario Hydro or the equivalent in your area) the correct location of the pole from which you will receive service. Your service head must be properly located with respect to the Hydro service pole.

(b)   **Type of Cap to Use** - Rule 6-114

There are a number of different caps permissible depending on the type of service conduit or cable you use. See under specific type.

Slip-on Type

(c)   **Height of Entrance Cap** - Rule 6-112

There are two things to watch for:

1st. .-   That you can obtain the minimum line clearances, given on page 14,

AND

2nd. -   Rule 6-116(2) says the entrance cap must be higher than the wire holder which supports the Hydro line. This rule says the cap must be between 150 mm and 300 mm (between 6 to 12 in.) above the Hydro line insulator as shown on page 15.

If, in your case, it is difficult to locate the cap above the line insulator you should check with your local Inspector first before you ignore this rule.

Some background Information for the student.

Rule 6-116(b) was originally designed to prevent water travelling down the Hydro service line to the splices at the entrance cap and there, at the splice, find its way into the conductor itself. Water in this location (actually around the conductor strands inside the conductor insulation) would be drawn along by capillary action between the conductor strands and on down to the meter base where it would collect and do all kinds of nasty things.

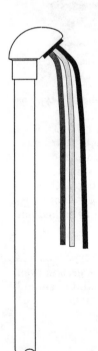

(d)   **Drip Loops** - Rule 6-302(3)

This Rule requires that you leave at least 750 mm (29.5 in.) of service conductor hanging out of the entrance cap.

This long length may seem wasteful and sometimes is, but that's what the rule requires. The purpose is to provide sufficient length for Hydro crews to form acceptable drip loops that will prevent water following the service conductors into the service conduit and down into the service equipment as described above.

(e)   **Location of Entrance Cap** - Rule 6-112(3)

This is an "out of reach" rule. The entrance cap must be placed so that **all** open conductors (conductors not in conduit) are above the window. If they are below or alongside the window, or if they run in front of the window, as shown below, they must be at least 1 m (39.37 in.) away from the window. This rule applies to all windows **even those which cannot be opened.**

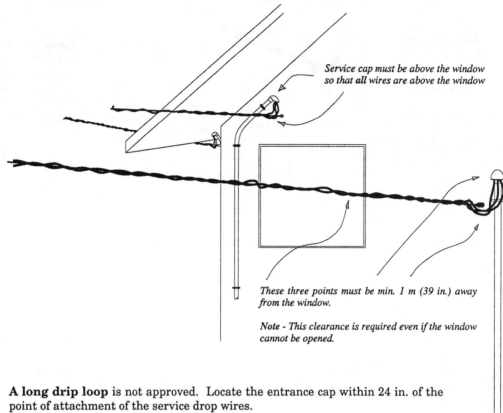

Service cap must be above the window so that all wires are above the window

These three points must be min. 1 m (39 in.) away from the window.

Note - This clearance is required even if the window cannot be opened.

Note 1   **A long drip loop** is not approved. Locate the entrance cap within 24 in. of the point of attachment of the service drop wires.

2   **Hydro Service drop leads** - These should terminate at the first contact with the building at the entrance cap location. However, where for one reason or another the entrance cap cannot be brought to that first contact point the rule will allow service conductors to be run along the outside wall of a building to the entrance cap location. It must be attached to line insulators spaced along the outside surface of the building. Special permission is not required but this arrangement should be used only in special cases and only after consultation with your local Inspector.

3   **Snow Slides** - Bulletin 6-2-0 (1)(e) - Many services have been pulled out or have been severely damaged by snow sliding off a roof during the winter months. It is important that the entrance cap be very carefully located so that such slides cannot harm the service conduit and the Hydro service lines. The gable end of the roof is the preferred location for the entrance cap.

In the illustration above, one example shows the service leads terminating on the wall below the roof overhang. This is a poor location and may not be acceptable to Hydro because snow sliding off the roof during the winter months could damage these lines - actually pull them out.

Note - Service masts which run through any part of a roof, particularly the lower part of a roof, **may** also not be acceptable to some utility companies. Therefore, in heavy snowfall districts the entrance cap should be located on the **gable end** or similar location on the house where it will not be subject to damage by snow slides.

9. SERVICE MAST REQUIREMENTS - Rule 6-112(4)(5)(6) & (8)

Where a service mast is needed to raise the entrance cap and service leads to the required height, the following should be complied with:

PIPE MAST - Surface mounted

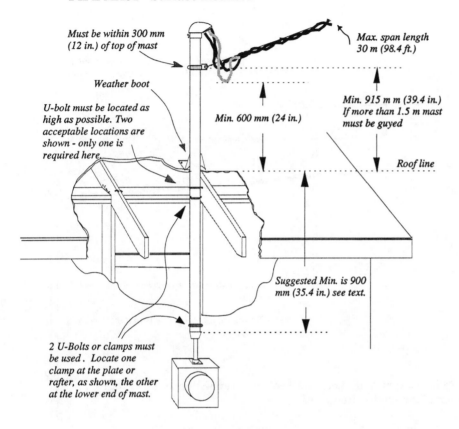

Must be within 300 mm (12 in.) of top of mast

Weather boot

U-bolt must be located as high as possible. Two acceptable locations are shown - only one is required here.

Min. 600 mm (24 in.)

Max. span length 30 m (98.4 ft.)

Min. 915 m m (39.4 in.) If more than 1.5 m mast must be guyed

Roof line

Suggested Min. is 900 mm (35.4 in.) see text.

2 U-Bolts or clamps must be used . Locate one clamp at the plate or rafter, as shown, the other at the lower end of mast.

Caution - if the mast is installed in the path of sliding snow from a sloping roof of smooth hard material such as plastic or metal, the Inspector may ask for an additional guy or brace to support the mast.

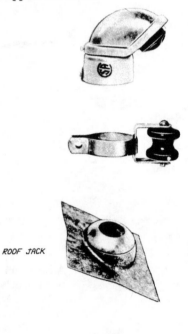

ROOF JACK

Parts for an Acceptable Mast

This may appear complicated but all the parts necessary for a service mast can be purchased in most building supply stores. The sales people in these stores will usually assist you in selecting the correct pieces for easy assembly of an acceptable service mast.

REDUCER IF REQUIRED

Note - If service mast kits are available from your supplier it is probably best to get one. That way you can be sure you have all the required parts. However, Rule 6-112(4) and the Code Appendix to this rule do not require that you use a certified electrical service mast kit. All that is required is that all the component parts which you will use are suitable for use in a service mast.

A CSA or other certified electrical service mast kit will have a length of steel pipe specially certified for service mast use; an entrance cap and a special fitting for the lower end to reduce pipe size to your service conduit size. It will also have the necessary U-bolts or clamps for fastening. There will be a roof boot and an insulated wire holder complete with pipe clamp so that it can be fastened to the mast as shown in the illustration above.

(a)   **Dimensions**

Pipe Mast   Use an approved mast kit or use 2.5 inch diameter rigid steel electrical conduit.
Conduit must be without joint to insure maximum strength.
Use suitable fittings at each end.

Wood Mast   Wood masts are not generally acceptable except by special permission.

(b)   **Above and below the roof line**

Minimum height of lowest drip loop above roof line is 600 mm (23.6 in.).

Minimum height above roof line for the Hydro service line attachment to mast is 915 mm (36 in.).

(c)   **Fastening - Mast Support**

The pipe mast must be securely fastened to the building structure with 3 - ½ inch U-bolts or with single bolt type fastener as shown in the illustration above.

Minimum distance between points of mast supports is not given by Code rules. This is determined on site and is dependent on the length of Hydro lines required to serve your building. The drawing shows 900 mm (35.4 in.). This distance is usually acceptable in most cases.

Maximum height of Hydro line attachment above roof line is 1.5 m (59 inches). Where the point of support must be more than 1.5 m above the roof line the rule says we need to install a guy line to provide additional support for the mast. Do not use clothes line for this purpose. A ¼ inch galvanized steel cable is usually acceptable. Use galvanized steel cable clamps and anchor point to complete the installation.

10 LENGTH OF SERVICE CONDUCTOR - Rule 6-206(1)(e), 6-208

According to these rules the service panel must he located "as close as practicable to the point where the service conductors enter the building". That's a good rule - keep it as short as possible for two good reasons.

(1)   Because it is an unprotected conductor - only the Hydro line fuse, which is ahead of the transformer, is protecting you, and

(2)   The service run is very costly.

(a)   Maximum length permitted OUTSIDE of the building

Where the service conductors are in a conduit or where they are in a service cable and run on the outside of the house the code does not limit its length. It may be any reasonable length. The high cost of the required material usually keeps this as short as possible.

(b)   Maximum length permitted INSIDE the building

The rule says . . . . . . through the wall only . . . no further.

Specifically the rule says service equipment shall be located "as close as practicable to the point where the consumers service conductors enter the building". This means that in most cases the service conduit or cable would run down the outside surface of the building to the meter base then through the wall into the back of the service panel. In this way the length of the service conductors actually inside the building would be only that portion which runs through the wall, usually less than 12 inches.

Note   The definition of "inside" the building is usually considered to be anywhere under the first layer of paint on the outside of the building. This means that service conductors in an outside combustible wall, in an attic or in a combustible crawl space are all considered to be **inside the building.**

(c)   Longer service runs - Where more length is needed to place the service equipment in a satisfactory location the Inspector may, by special permission, permit a longer run inside the building.

Where the panel location is well within the building we can take advantage of another rule which permits us to run service conduit or cable inside a building if it is enclosed in at least 2 inches of solid concrete (all around covering) as shown below, or if it is **buried in the ground** under the floor.

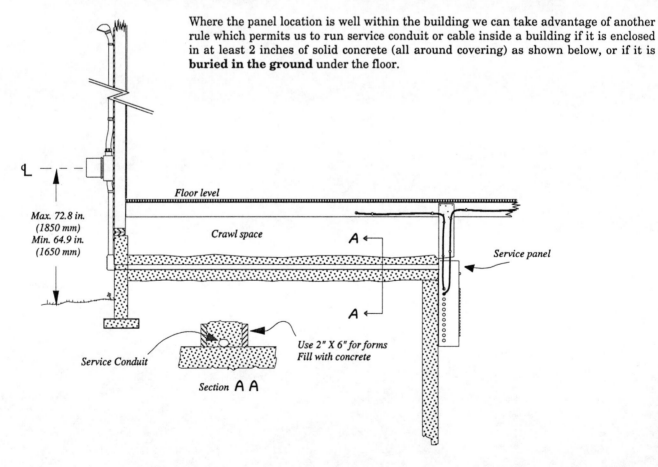

Max. 72.8 in.
(1850 mm)
Min. 64.9 in.
(1650 mm)

Floor level

Crawl space

A

A

Service panel

Service Conduit

Use 2" X 6" for forms
Fill with concrete

Section A A

11   SERVICE CONDUIT or SERVICE CABLE

There are several different wiring methods permitted for service conductors.

(a)   **EMT - Thinwall Conduit** - is **not permitted** for service installations in Ontario.

(b)   **Rigid Metal Conduit** - Rule 6-302(1) This is very good but very costly and difficult to work with.

(c)   **Rigid PVC Conduit** - Rule 6-302(1)(a) In most cases PVC conduit is acceptable for services. It has an advantage - very few tools are required to install it.

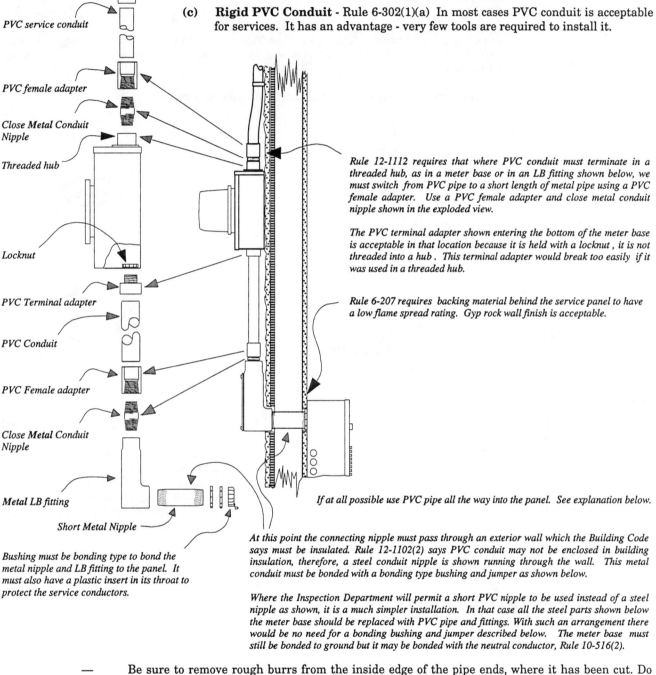

PVC Entrance Cap

PVC service conduit

PVC female adapter

Close Metal Conduit Nipple

Threaded hub

Locknut

PVC Terminal adapter

PVC Conduit

PVC Female adapter

Close Metal Conduit Nipple

Metal LB fitting

Short Metal Nipple

Bushing must be bonding type to bond the metal nipple and LB fitting to the panel. It must also have a plastic insert in its throat to protect the service conductors.

Rule 12-1112 requires that where PVC conduit must terminate in a threaded hub, as in a meter base or in an LB fitting shown below, we must switch from PVC pipe to a short length of metal pipe using a PVC female adapter. Use a PVC female adapter and close metal conduit nipple shown in the exploded view.

The PVC terminal adapter shown entering the bottom of the meter base is acceptable in that location because it is held with a locknut , it is not threaded into a hub . This terminal adapter would break too easily if it was used in a threaded hub.

Rule 6-207 requires backing material behind the service panel to have a low flame spread rating. Gyp rock wall finish is acceptable.

If at all possible use PVC pipe all the way into the panel. See explanation below.

At this point the connecting nipple must pass through an exterior wall which the Building Code says must be insulated. Rule 12-1102(2) says PVC conduit may not be enclosed in building insulation, therefore, a steel conduit nipple is shown running through the wall. This metal conduit must be bonded with a bonding type bushing and jumper as shown below.

Where the Inspection Department will permit a short PVC nipple to be used instead of a steel nipple as shown, it is a much simpler installation. In that case all the steel parts shown below the meter base should be replaced with PVC pipe and fittings. With such an arrangement there would be no need for a bonding bushing and jumper described below.   The meter base must still be bonded to ground but it may be bonded with the neutral conductor, Rule 10-516(2).

—   Be sure to remove rough burrs from the inside edge of the pipe ends, where it has been cut. Do this with any knife.

—   Be sure the pipe and fittings are clean and dry before applying solvent cement.

Note -   The illustration shows a metal pipe nipple running through an exterior wall, which is filled with building insulation, then into the back of the service panel. This metal pipe must be bonded to ground with a bonding bushing and jumper as described below under "Size of Bonding Jumper".

It would be much better and very much simpler to use PVC pipe and fittings all the way into the panel except that the very strict interpretation of Rule 12-1102(2) requires metal pipe where it must run through building insulation. This rule was not intended to apply where there is no need to do so, as is the case with single family residential services. The fact is, this rule is badly written and the Appendix on page 581 in the Code seems to confirm this. Until the wording of this rule is changed it is important that you know it is there and that it could be correctly applied in this case. Check with your local Inspector if PVC pipe is acceptable in this location.

If a PVC pipe nipple is acceptable in this location don't even bother reading all that stuff about bonding bushings and bonding jumpers below. The rules require those silly things only where metal pipe is used.

Size of Bonding jumper - This is based on Table 41 in the Code. This table requires a #8 bare copper jumper for a 100 ampere service or a #6 bare copper jumper for any service rated 101 to 399 amperes.

Two Locknuts are required in addition to the bonding jumpers.

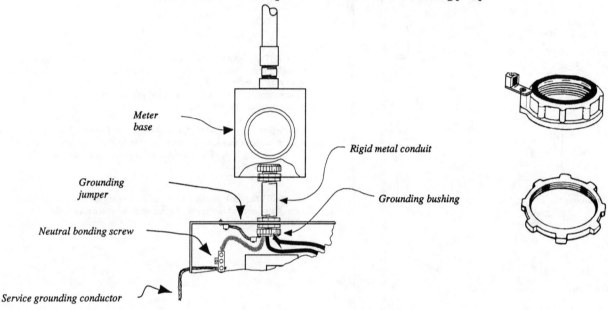

Meter base

Rigid metal conduit

Grounding jumper

Grounding bushing

Neutral bonding screw

Service grounding conductor

Locknuts are dished, that is they are not flat. They are designed to bite through the paint or rust and into the metal of the box.

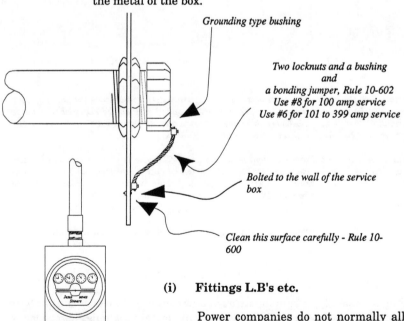

Grounding type bushing

Two locknuts and a bushing and a bonding jumper, Rule 10-602 Use #8 for 100 amp service Use #6 for 101 to 399 amp service

Bolted to the wall of the service box

Clean this surface carefully - Rule 10-600

L.B. Fitting

Bushings must be bonding type as shown and where the service conductors are larger than #10 the bushing must have an insulating ring mounted in its throat. This ring is required by Rule 12-906 to protect conductors from abrasion.

If you can find a way to use a PVC pipe nipple instead of the steel conduit nipple you do not need to fuss with the silly bonding jumper as shown. As noted above PVC pipe may not be enclosed in building insulation but if you can find a way to prevent that happening in an outside insulated wall, to the satisfaction of the electrical Inspector, the installation would be a lot easier.

(i)   Fittings L.B's etc.

Power companies do not normally allow conduit fittings to be installed ahead of the meter base, however, on the load side of the base you may install as many fittings as you require. Where it cannot be avoided and a fitting must be installed ahead of the base, you should seek Hydro permission first before proceeding.

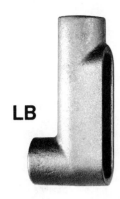

LB

Problems using L.B. fittings - Avoid using these fittings where service conductors are larger than #3 copper. Great care is needed when seating, or forming conductors into the fitting. Pull them into place one at a time being careful not to damage the insulation. It's okay to use a hammer to form the conductors into the fitting but do not apply directly on the insulation. Use a smooth piece of wood against the conductors and drive that gently with a hammer.

Double L.B. Problem - Never use two LB fittings back to back for any conductors larger than #3. It is very difficult to force the conductors into the fitting without damaging the insulation in the process. Replace one of the fittings with a 90° bend.

L.B. Fittings must be Accessible - These must be located where they will remain accessible for any maintenance work that may be required in the future.

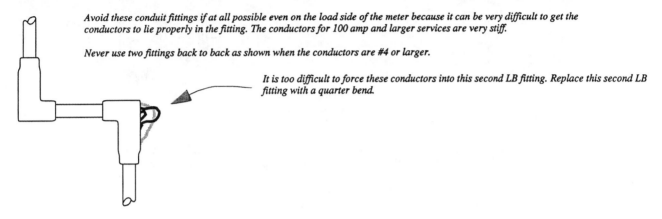

Avoid these conduit fittings if at all possible even on the load side of the meter because it can be very difficult to get the conductors to lie properly in the fitting. The conductors for 100 amp and larger services are very stiff.

Never use two fittings back to back as shown when the conductors are #4 or larger.

It is too difficult to force these conductors into this second LB fitting. Replace this second LB fitting with a quarter bend.

(ii)   Bends - Rules 12-1108 & 12-1112

Manufactured Bends - The maximum number of bends permitted by the rule is the equivalent of four quarter bends (4 - 90° bends or any combination of bends where the total does not exceed 360°).

Home Made Bends - Not Recommended - Avoid this if you can but if your installation requires a special bend you can take advantage of the fact that PVC conduit may be bent in the field. To do this the pipe must be carefully heated to 260° F at the location of the proposed bend. It is best to use a heat gun to heat the pipe. An open flame could also be used, the Code does not prohibit it, but in that case it must be done **very carefully to avoid damage.** Too much heat will char the pipe or blister its smooth surface and that could be grounds for rejection. Make sure the bend section is uniformly heated all around for a distance of about 10 times the pipe diameter before you attempt to bend it. Improper heating or improper bending procedures may **cause the pipe to kink or collapse and that would very likely result in a rejection** of your installation. One more thing, there is a small amount of spring back when the pipe cools. To compensate for this loss you will need to overbend it a few degrees more than is required.

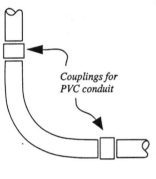

Couplings for PVC conduit

(iii)   Strapping - Rule 12-1114

Install one strap at the top near the entrance cap, another near the meter base. Install additional straps as required so that the maximum distance between straps is not more than 2 m (approx 6 ft.).

Note that PVC conduit has a high coefficient of expansion. This means that it is a bit longer in the summer than in the winter. If it is exposed to direct sunlight it is even longer. For short runs, 10 ft or less, this is not a great concern but longer runs should be supported with straps that will permit the conduit to slide when it expands with temperature change.

(iv)   Sealing - to stop breathing - Rule 6-312

The seal required by this rule prevents the warmer inside air from escaping through the service conduit. It has been found that warm air, if allowed to flow through the service conduit, will condense to water in sufficient quantity to damage service equipment.

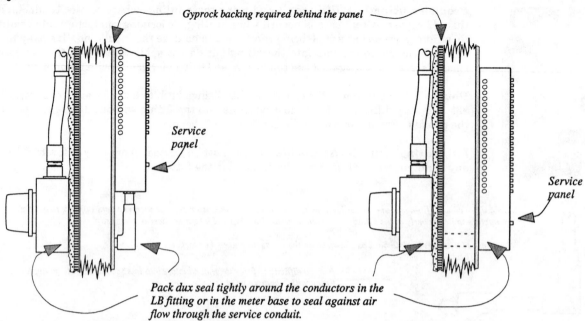

Gyprock backing required behind the panel

Service panel

Service panel

Pack dux seal tightly around the conductors in the LB fitting or in the meter base to seal against air flow through the service conduit.

The seal is usually made with a soft putty like substance called DUX SEAL. To be most effective DUX SEAL is carefully placed to close all openings around the conductors in the last opening before the conduit leaves the inside of the house or in the first opening outside of the house. Only one seal is required. Make sure all openings around each conductor are completely sealed to prevent any air passage through the conduit.

(v)   **Holes in Outer Walls, floors or Roofs** - Rule 12-018, 12-926 & Bulletin 12-1-6 - These rules require that we fill in any openings around conduit or cable where these pass through an outer wall or a roof or a floor. This is not exactly electrical work but your Electrical Inspector will check this part of your installation and he is required to yell at you, or stomp his feet, if this is not properly done for the final inspection.

(c) **Service Cable - From an Overhead Hydro Supply**

    (i)   **Type** - Rule 6-302

        Type TECK 90 cable is permitted for service work. It is easier to install than conduit.

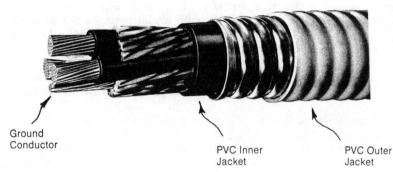

Alcan multi-conductor
Teck-90 Cable

Ground
Conductor

PVC Inner
Jacket

PVC Outer
Jacket

    (ii)   **Size**

        The ampacity required for a cable is the same as an equivalent size conductor in conduit. See pages 6 & 7 for sizes required.

    (iii)   **Type of Entrance Cap to Use** - Rule 6-114

        Heat Shrink - a kind of plastic sock that slips over the top end of your service cable. To install this sock you must remove the outer PVC jacket, the armour and the inner PVC jacket for approx 30 inches. You should now have 30 inches of insulated conductors exposed for the drip loops and connection to Hydro lines.

        Slip this sock in place, then very carefully apply heat, as evenly as possible, with an open gas torch or heat gun, (heat lamps could also be used) as long as the heat applied is approximately 250°F. Apply heat evenly and do not overheat the material. When heat is applied the sock will shrink to fit snugly in place. It will seal it from rain but remember, the rule says it must still face downward when installed. **Be sure to allow enough length to do this.**

    (iv)   **Cable Connectors**

        Use weatherproof connectors where the cable connects to the top of the meter base. Be sure to use the correct size and type for the cable you are using. Your supplier will advise you on this.

        Dry type connectors may be used where they are not exposed to the weather.

        Note - Anti-Short Bushings, shown at left, must be used with dry type connectors. This is a fibre or plastic bushing that fits into the end, inside the armour of the cable. It protects the conductor insulation at the point where they issue from the armour. This thing is wholly inside the connector but its presence can be easily verified through the small openings provided for this detective work by the manufacturer. It is easier to put it in place when the cable is being installed. Later, after it has been rejected, it is more difficult to do this.

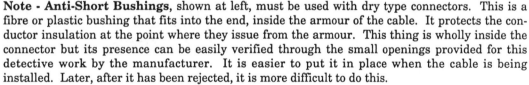

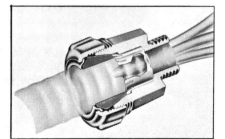

(v)   **Length of Service Cable Permitted** - Rule 6-206 & 6-208

Outside of the house - Service cable which is run along the outside surface of the house may be as long as it needs to be to get where your going. However, this cable is expensive; you will want to keep it as short as possible.

Service cable run INSIDE the house - The rule says service equipment, (the service panel) must be located so that this part of the run can be kept as short as practicable. This usually means through the wall and just a teensy bit bit more - a few feet perhaps. This part of the installation will normally be with a metal conduit nipple not with TECK cable because, in this location, it is much easier to install.

Longer service runs - Where more length is needed to place the service equipment in a satisfactory location the Inspector may, by special permission, permit a longer run inside the building.

Very Long Service Runs - Where the panel location is well within the building we can take advantage of another rule which permits us to install long runs of service cable inside a building if it is enclosed in at least 2 inches of solid concrete (all around covering) or if it is buried in the ground under the floor. The illustration on page 20 shows such an installation for conduit. A cable would be installed a little differently but the principle remains the same; any conduit or cable which is enclosed in at least 2 inches of concrete is considered to be outside the house and may, therefore, be as long as needed.

(vi)   **Strapping** - Rule 12-706

Cable must be strapped every 78 inches (2 m).

(vii)   **Meter Connections** - Rules 10-516(2) & 10-906(3)

Do not cut away the PVC jacket on this cable, it must run into the connector in this wet location.

The bare bonding wire must terminate in a separate grounding lug which is bolted to the side wall of the meter box, as shown. Do not use one of the wood screws for this purpose, Rule 10-906.

Cable neutral must be spliced in the meter base.

Stuff dux seal around conductors at this point.

(viii)   **Mechanical Damage** -.Greater care is needed when installing this cable than is required when installing conduit. This cable is more easily damaged with driven nails or where it is run on the surface of a wall.

In the illustration below, the cable is shown running through the plate and over the broken edge of the foundation wall. This section of cable (where it runs through the plate) is subject to damage and must be protected. Use heavy gauge metal plates to protect the cable at these points. The side plates of metal outlet boxes do this very well.

Where the cable is run on the surface of the wall and where it may be subject to mechanical damage (in locations such as a garage or carport for example) the cable may be protected with wooden or metal guards or a short section of metal pipe may be used.

The illustration below shows TECK cable being used above the meter base to an overhead supply and from the meter base to the panel. It also shows PVC conduit from the meter base downward to an underground Hydro system. Obviously, only one of these systems will be available - either overhead supply or underground supply and therefore only the TECK cable running up or the PVC conduit running down is necessary. Both systems are shown to indicate different methods of installation.

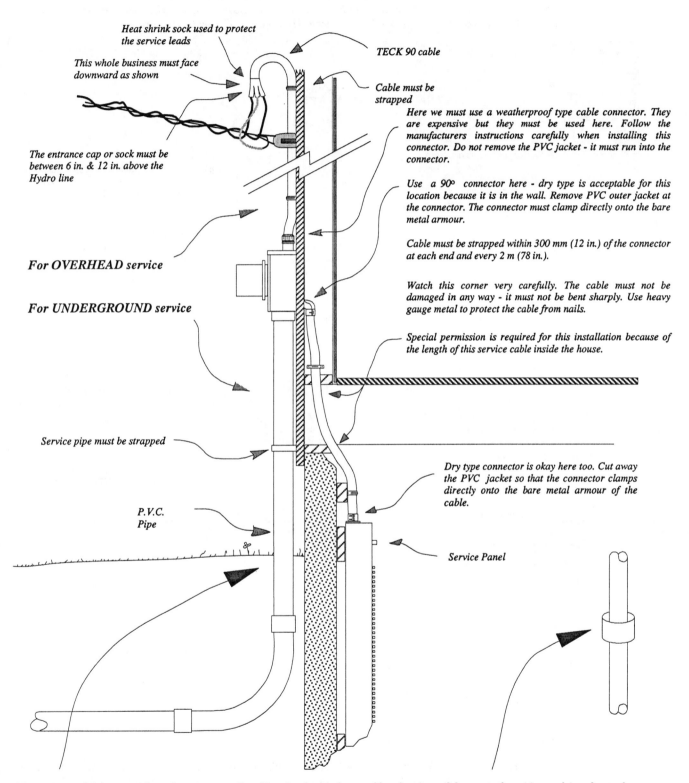

Heat shrink sock used to protect the service leads

This whole business must face downward as shown

TECK 90 cable

Cable must be strapped

Here we must use a weatherproof type cable connector. They are expensive but they must be used here. Follow the manufacturers instructions carefully when installing this connector. Do not remove the PVC jacket - it must run into the connector.

The entrance cap or sock must be between 6 in. & 12 in. above the Hydro line

Use a 90° connector here - dry type is acceptable for this location because it is in the wall. Remove PVC outer jacket at the connector. The connector must clamp directly onto the bare metal armour.

Cable must be strapped within 300 mm (12 in.) of the connector at each end and every 2 m (78 in.).

For OVERHEAD service

For UNDERGROUND service

Watch this corner very carefully. The cable must not be damaged in any way - it must not be bent sharply. Use heavy gauge metal to protect the cable from nails.

Special permission is required for this installation because of the length of this service cable inside the house.

Service pipe must be strapped

Dry type connector is okay here too. Cut away the PVC jacket so that the connector clamps directly onto the bare metal armour of the cable.

P.V.C. Pipe

Service Panel

If the service conduit must pass through a concrete walk at this point there is danger of frost heaving and damage to the service conduit and meter base.

To prevent this damage install a short length of larger pipe, or wood blockout, where the service conduit runs through the sidewalk. This will permit the concrete sidewalk to move up and down without pulling the meter base off the wall.

12 SERVICE from an UNDERGROUND SUPPLY

(a)   **Hydro Connection** - Contact Hydro for location of trench termination at the property line.

(b)   **Service Feeder to House** - For the underground portion from the property line to the house there are at least two choices as far as the rules are concerned. These are as follows: Conductors may be installed in conduit as shown in the illustration on page 27, or as shown below, service cable certified for direct burial without conduit may be used. Check with your local Hydro office if they have a preference.

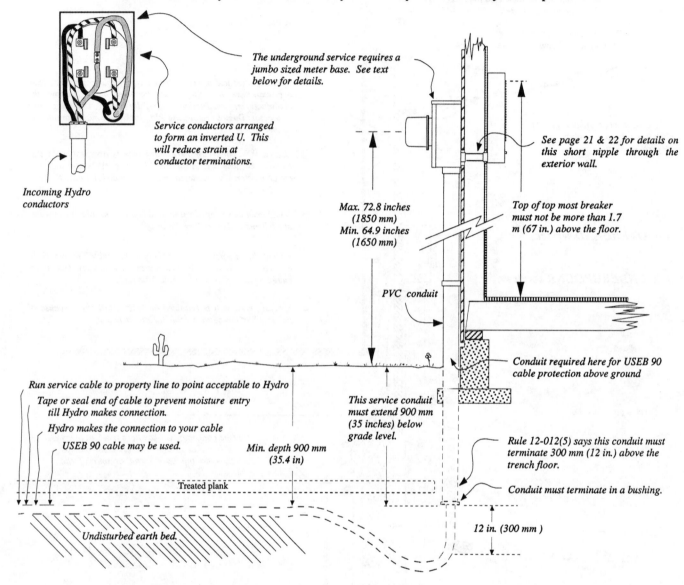

The underground service requires a jumbo sized meter base. See text below for details.

Service conductors arranged to form an inverted U. This will reduce strain at conductor terminations.

Incoming Hydro conductors

See page 21 & 22 for details on this short nipple through the exterior wall.

Max. 72.8 inches (1850 mm) Min. 64.9 inches (1650 mm)

Top of top most breaker must not be more than 1.7 m (67 in.) above the floor.

PVC conduit

Conduit required here for USEB 90 cable protection above ground

Run service cable to property line to point acceptable to Hydro

Tape or seal end of cable to prevent moisture entry till Hydro makes connection.

Hydro makes the connection to your cable

USEB 90 cable may be used.

This service conduit must extend 900 mm (35 inches) below grade level.

Min. depth 900 mm (35.4 in)

Rule 12-012(5) says this conduit must terminate 300 mm (12 in.) above the trench floor.

Conduit must terminate in a bushing.

Treated plank

Undisturbed earth bed.

12 in. (300 mm)

(c)   **Size of service conduit** - Building Code Rule 9.35.5.6.(1)

The Building Code requires that this vertical section of conduit from the trench to the meter base must be at least 2 inches in diameter. However, bear in mind that each of the separate power utilities in Ontario may set its own requirements for this service conduit. They may require this conduit to be larger than 2 inches but it may not be smaller than 2 inches.

(d)   **Depth of burial** - Electrical Code Rule 12-012(5) and Building Code Rule 9.34.4.5.(1)

End of Service Conduit - The Building Code requires this service conduit to extend at least 900 mm (35 in.) into the trench and Rule 12-012(5) requires this conduit to terminate 300 mm (12 in.) above the trench floor. That straight section of cable from the end of the service conduit to the trench floor is provided to permit movement as a result of frost heaving.

Depth of Service Cable - Check with your local Hydro office for minimum depth of service cable burial. Each utility will have its own standards. Some utilities will provide and install this cable in a trench you have provided others may not provide or install it. In some cases you will be required to install Big O pipe for the service cable. This pipe is not acceptable as a service conduit, it serves only as a chase for quick removal of the service cable in the event of trouble or upgrade. For this reason, when service cable is drawn into Big O pipe it is considered to be direct buried, therefore, the cable must be suitable for direct burial. Cable such as USEB is acceptable for this service.

(e)   Meter base - for a service supplied from an Underground Distribution System

Minimum size meter base permitted is 200 amps. This base must also have the following minimum dimensions: $17\frac{1}{2}$ inches high, $10\frac{1}{8}$ inches wide, $4\frac{1}{2}$ inches deep. This base is often referred to as the Jumbo size meter base. The reason for this requirement is twofold. First, Hydro needs the larger base to provide adequate physical space when installing their large service cables. Secondly, the supply conductor size they use for these services is often too large for the terminals in a 100 amp base.

(f)   Sealing - Rule 6-312

The seal required by this rule prevents the warmer inside air from escaping through the service conduit. It has been found that the warm air, if allowed to flow, condenses to water in sufficient quantity to damage the service equipment.

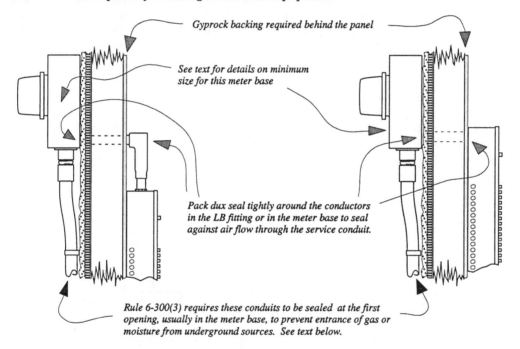

Gyprock backing required behind the panel

See text for details on minimum size for this meter base

Pack dux seal tightly around the conductors in the LB fitting or in the meter base to seal against air flow through the service conduit.

Rule 6-300(3) requires these conduits to be sealed at the first opening, usually in the meter base, to prevent entrance of gas or moisture from underground sources. See text below.

This seal is usually made with a soft, putty-like substance called DUX SEAL. To be most effective the DUX SEAL is placed around the conductors in the last opening before the conduit leaves the warmer area.

Underground service conduit from an underground distribution system - Rule 6-300(3) requires service conduit to be sealed at the first opening above ground. This is to prevent stray gas or moisture entering the enclosure, (the meter base). Dux seal used to prevent breathing through the conduit nipple which runs through the wall can also be used here to seal against gas entering the meter base. Where the meter base is below the level of an underground service piping system from the property line Dux seal is not a sufficient barrier to prevent water entry. Check with Hydro office for advice on an acceptable seal.

Conduit Entry into a House from an underground distribution system - Rule 6-300(4) requires the underground service conduit to enter a house above finished grade. It may not enter directly below ground. There is a good reason for this requirement. Gas leaking from a nearby underground gas line could travel along the electrical pipe and enter the house. The volume of gas entering would likely not amount to much but if it became trapped and began to accumulate it would only need a tiny spark to cause a violent explosion. Where the conduit emerges outside of the house explosive gas vapours can evaporate safely to the atmosphere.

13 METER BASE INSTALLATION

(a)   **Types**

In general there are two types of meter bases available, the round and the square or shoe box type. Many of the power utilities will not accept the round base for any size installation. Some utilities require what is called the "jumbo" size base. Check first with your local utility office.

Note 1   For an underground service use a 200 amp. meter base with the minimum dimensions given on page 29. This large base must be used to provide adequate working space for Hydro crew to do their work inside the base.

Note 2   If the knockouts provided by the manufacturer of the base are not in the correct position and new holes need to be punched out, these new holes must be totally below any live parts in the meter base. Only where a meter base is installed totally indoors, as in a service room in an apartment building, may new holes be cut above the live parts in the base.

By the way, if you plan to use an old round type meter base which has side or back conduit entry holes, be careful. First, because most utilities do not accept them and second, only the top and bottom entries in round bases may be used where these round bases are allowed. One more thing, many of those old meter bases were not designed to permit splicing of the neutral in the base. This means that it may not be suitable if you are using TECK cable for service conductors.

(b)   **Locations** - Bulletin 6-1-4 and Ontario Building Code Rule 9.34.4.2.(2)

The meter base must be located within 3 m (118 inches.) from the front of the house.
The front of the house is the side nearest the Utility distribution line.

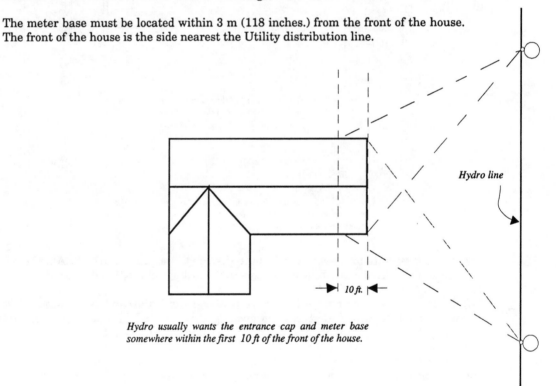

Hydro line

Hydro usually wants the entrance cap and meter base somewhere within the first 10 ft of the front of the house.

Meter locations must be carefully chosen. Some of the things to watch for are:

(i)   **Carports**

If you plan to face your meter base into the carport you should know that there is no code rule or bulletin which specifically says you may not have it there. However, before you install it in the carport you should talk to your local Hydro people. They may not like it there.

Fact is, no matter how careful you are as a driver, when you are backing in your 24 ft. Winnebago that meter is subject to damage if it faces into the carport.

Every year there are thousands of carports closed in to make a garage or an additional bedroom. If the meter faces into the carport the space cannot be closed in to convert it into a bedroom without first relocating the meter base. This is usually a costly relocation.

(ii)   Porch

If it is an open porch it may be an acceptable location now but remember you may want to close it in later on in the future. A closed in porch is a heat saver in the winter time, so avoid the hassle, follow old Chinese proverb - don't do it in the porch.

(c)   Meter Base Height - Ontario Building Code Rule 9.34.4.2.(1)

Meters shall be located on an outside wall, facing out. They must be located somewhere between 64.9 in (1650 mm) and 72.8 in (1850 mm) above finished grade. All heights to be measured from the center of the base to permanent grade level.

(d)   Connections

Rule 10-516(2) requires the neutral to be connected to the meter base as shown. All modern meter bases have provision for connecting the neutral from both the line and the load. Some of the older meter bases may not be properly equipped to make a splice in the neutral conductor. In those cases simply bare a section of the conductor where it passes the bonding terminal in the meter base - then slip it into the lug provided and tighten. Do not cut this conductor unless it is necessary to do so.

240 Volts 3-wire

Incoming 2 black wires (or a black and a red wire) from Hydro connect to the top two terminals.

2 Black Load wires (or a black and a red wire) connect to the lower terminals.

The white wire is also used to bond the meter base, Rule 10-516(2). Connect this neutral conductor to the bonding terminal in the base as shown.

If you are using an older round base check the bonding terminal. In some of old bases the bonding terminal was not designed for splicing the neutral conductor. If it has only a single termination point do not cut and splice the neutral - simply bare a section and lay it in the connector, then tighten as required.

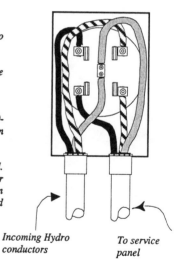

Incoming Hydro conductors   *To service panel*

(e)   Support

Support the meter base with wood screws through the two or more factory drilled holes in the back of the base. Make sure it is in a reasonably accurate upright position.

(f)   Blank Cover

In some districts the power utility will energize the electrical service but not install a meter until several days later. During this time a blank cover is required to prevent anyone coming in contact with live parts. A disk of ¼ in. plywood may be used for this purpose.

(g)   **Sealing Rings**

Meter sockets are equipped with either a screw type sealing ring or a spring type ring. Some Power Companys will not accept the spring clip type ring. Check with your power company.

(h)   **Fittings & LBs etc.**

Ahead of Meter - Some power companies do not allow fittings to be installed ahead of a meter base. This has something to do snitching power which power companys have decided is bad for business. However, on the load side of the base you may install as many as you require. Where it cannot be avoided and a fitting must be installed ahead of the base you should seek Hydro permission before proceeding. They may be satisfied if the fitting is within sight of a meter reader and the fitting has been drilled to permit Hydro personnel to install a seal. See also the note on page 23 regarding two LB fittings.

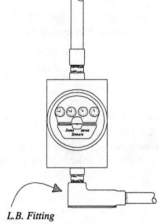

L.B. Fitting

14 SERVICE PANEL

(a) **Type - Fuse or Breaker Panel**

Both are acceptable, however, only the circuit breaker panel is in common use today. For this reason we will deal with circuit breaker panels only.

Typical Breaker Type Service Panel

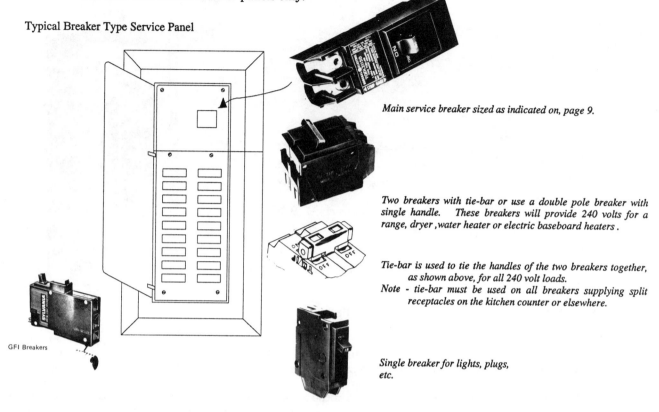

Main service breaker sized as indicated on, page 9.

Two breakers with tie-bar or use a double pole breaker with single handle. These breakers will provide 240 volts for a range, dryer ,water heater or electric baseboard heaters .

Tie-bar is used to tie the handles of the two breakers together, as shown above, for all 240 volt loads.
Note - tie-bar must be used on all breakers supplying split receptacles on the kitchen counter or elsewhere.

GFI Breakers

Single breaker for lights, plugs, etc.

NOTES - Rule 14-010(b) A revised rule.

1 **3-Wire Cable Entering the Service Panel** - If the cable supplies **split receptacles** anywhere in the building, the circuit breakers used to supply the three wire cable must be either two-pole type or two single pole breakers mounted side by side and their operating handles connected together with a tie bar as shown in the illustration above. There is a good reason for this requirement. In the case of split receptacles, both hot wires are connected to the same duplex receptacle and therefore, both must be disconnected before anyone can safely work on this device.

Where the three wire cables supply **ordinary duplex plug outlets or lights** and **each** is connected to only **one hot wire and the neutral** the circuit breakers supplying this three wire cable do not need to be two-pole type nor do they need to have a tie-bar connecting their handles.

Very briefly, 2-pole breakers or tie-bars are required for all breakers serving:

> *Any device, which is connected to both hot wires, such as a range, dryer etc.; and*
> *Any split duplex receptacle.*

All other breakers supplying 3-wire cables are not required to be 2-pole or be equipped with tie-bars.

2 **Bathroom Razor Outlet** - Rule 26-700(13) - These must be supplied with a "ground fault interrupter" type circuit breaker or you may use a ground fault interrupter type receptacle. The special transformer type razor outlet is no longer approved and may not be installed in new construction but may only be installed to replace a faulty unit in an existing installation.

(b) **Identify Circuits** - Kinds of Loads Served - Rule 2-100(2)&(3) and Bulletin 2-5-1.

Use a felt pen or some other permanent manner of marking next to the circuit breaker or fill in the circuit directory card provided by the panel manufacturer. The identification should look something like this.

IDENTIFY LOADS HERE NEXT
TO THE FUSE OR BREAKER.

EXAMPLES

RANGE	- 40 AMPS.
DRYER	- 30 AMPS.
WATER HEATER	- 20 AMPS.
ELECT HEATERS BASEMENT	- 15 AMPS.
LR & KIT	- 20 AMPS.
BEDROOMS	- 20 AMPS.
COUNTER PLUG	- 15 AMPS.

BASEMENT OR ATTIC SUITE PANEL - Rule 26-440 & 26-704(1)

Rule 26-440 has been changed again. It is now much less threatening and less costly to comply with. Where a single family house is being divided into two or more dwelling units the rule now says that all outlets in the two or more dwelling units can be served from one panel.

If you decide to have a separate panel in the suite, or elsewhere, it is not now required that all outlets in that suite must be served from that panel, any of the outlets in either suite may be served from either panel. This is good news because it means that existing outlets also need not be rewired to the new panel.

Caution - If the separate panel noted above **is separately metered,** then all the full wrath of the rules will be applied. It is not that Hydro dislikes the extra meter but that you now have a duplex or triplex and all the electrical rules and all the building rules for such buildings must be applied.

(c) **How Many Circuits I Need?**

The number of circuits needed for a given house is determined by the minimum SERVICE AMPACITY as shown on the Service Size Table on page 9 - - - AND - - - the number of outlets installed in that house.

The Table on page 9 has been designed to simplify this problem. This table specifies the required ampere rating of the service and indicates which list of materials, given on page 6, should be used for that particular house. Each list of materials also indicates the size of panel required.

There are two steps involved - proceed as follows:

> **Step 1** Determine minimum number of circuits required from the Table on page 9.
>
> **Note** The table on page 9 gives the minimum service ampere rating required. Next to it is a letter in brackets. This letter refers to the list of service material required on page 6. This list also indicates the minimum number of circuits required for that house.
>
> **Step 2** Complete the chart on page 35 to determine the actual number of circuits needed to supply the outlets you plan to install. Carefully fill in the chart to make sure you do not run short of circuits when the loads are finally being connected.
>
> **Result** The size of the branch circuit panel must be equal to step 1 or step 2, **whichever is greater.**

Step 3 MINIMUM CIRCUITS REQUIRED - Before doing this, see page 48.

Light Outlets .... count all light outlets. indoor and outdoor, (do not

count switch outlets) .... _____

Convenience Plug Outlets - Locate these plug outlets so that no point along the floor line is
more than 6ft. (1800 mm) from a plug outlet. This refers to:

Living room plugs .... _____

Family room plugs .... _____

Bedrooms plugs .... _____

Dining room plugs .... _____

Any Other Rooms or areas .... _____

Each Bathroom........... Minimum 1 plug receptacle required .... _____

Bathroom fan .... _____

Each Washroom.......... Minimum 1plug receptacle required .... _____

Each Hallway .... Minimum 1 plug receptacle required for each

hallway. See page 65 .... _____

Each Kitchen fan........ Counts as one outlet See page 49 .... _____

TOTAL **OUTLETS** REQUIRED = ... _____

then =. $\dfrac{\text{Total \textbf{outlets} required}}{12}$ = **Circuits** required .... _____

ADDITIONAL CIRCUITS REQUIRED

Outdoor plug outlets ....Minimum 1 circuit required, See page 75 .... _____

Carport or Garage..............Minimum 1 circuit required in each, See page 77 & 78 .... _____

Laundry room or area.......Minimum 1 circuit required, See page 74 .... _____

Kitchen ....Minimum 1 circuit for fridge................................. _____

Plus counter outlets, See page 70 .... _____

Larger Appliances ....Range - 2 circuits required .... _____

2nd. Range 2 circuits required .... _____

Dryer - 2 circuits required .... _____

Dishwasher - 1 circuit required................................ _____

Garburator - 1 circuit required................................ _____

Compactor - 1 circuit required................................ _____

Micro~wave oven - 1 circuit required.......................... _____

Instant Hot Water Heater - 1 circuit required .... _____

Hydromassage Bath-tub - 1 GFCI circuit required.............. _____

Built-in vacuum Cleaner - 1 circuit required .... _____

Furnace (gas or oil) - 1 circuit required .... _____

Boiler - 1 circuit required..................................... _____

Domestic Use Water Heater - 2 circuits required .... _____

Swimming pool - Motor Load. 1 circuit required .... _____

Lighting load - 1 circuit required .... _____

Freezer plug (Separate circuit not required but is better)............... _____

Sauna Heater - 2 circuits required .... _____

Hot tub motor load - 1 circuit required .... _____

Electric heater - 2 circuits required .... _____

Domestic Water Pump - 1 circuit required (check pump rating)........... _____

Any Special plugs or lights .... _____

Plus 2 spare circuits (Rule 8-108(2))... _____

TOTAL CIRCUITS REQUIRED   = _____

(d)   **Sub-feeder to 2nd panel**

Sometimes it is an advantage to install a 2nd panel to supply the electrical loads in specific areas such as a kitchen or a garage. The kitchen requires a number of separate circuits for special loads such as the fridge, micro-wave oven, compactor etc.. The basement area directly below the kitchen is usually a good location for that second panel. Remember, the code does not require it. All the loads in the house may be served from the main service panel, however, this may require a very large service panel and a lot of costly long home runs. For this reason it may be an advantage to install a second panel - it may be a saving.

Don't forget, this second panel must be located with the same care and attention you used to locate the main service panel. All the rules regarding panel location, height, accessibility etc. as outlined below for a main service panel, must also be applied to the second panel. It may not be put just anywhere.

Service panel   *Sub-panel*

SIZE of PANEL and FEEDER CABLE REQUIRED

Section 8 of the code has nothing to say about minimum sizes for a 2nd panel in a single family dwelling. We cannot apply Rule 8-200 to a sub-panel because that rule must be applied to the whole house, not just to a part of the load, and it sets the minimum size at 60 amps. This rule cannot, therefore, be properly applied to our sub-feeder.

However, all is not lost. We may apply a thumb rule to arrive at satisfactory sizes. The "Thumb Rule" goes like this:

For any size house up to approximately 4000 sq. ft. floor area the following is usually acceptable.

EXAMPLE 1

Lighting Loads Only - If the 2nd panel will supply only lighting loads:

Sub-feeder Size   #10 copper loomex cable. 30 amp fuses or circuit breakers in the main panel.

Sub-panel Size   8 or 12 circuit panel. We may have as many as we wish. This is usually governed by the number of outlets per circuit, the area served and the bank account. Make sure your panel is large enough for the present load and for some future load additions.

EXAMPLE 2

Kitchen Electrical loads - If the 2nd panel supplies lights and plugs in the kitchen area as well as other loads such as the garburator, dishwasher compactor etc. but not an electric range or dryer or any electric heating:

Sub-feeder Size   #8 copper loomex cable. 50 amp fuses or circuit breakers.

Sub-panel Size   12 or more circuit panel is recommended. It is better to have too many circuit spaces for what you need now than to have too few spaces for your loads now and nothing for future load additions.

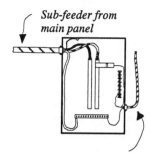

Sub-feeder from main panel

Only one branch circuit shown

Connect the black & red wires to the bus terminals.

The white wire connects to the neutral bus.
There should be no connection between the neutral and the metal enclosure - if there is a bonding screw, or jumper, remove it.

The bare wire connects to the enclosure at each end. This is an important connection - all grounding for the branch circuits depend on these connections.

Note 1   The range, dryer, and electric heating loads are not normally supplied from the 2nd panel. These loads are usually supplied directly from the main service panel.

Note 2   For a private Garage Panel - see under "Garage Wiring" on page 111.

Note 3   Additions & Renovations to an Existing House - See page 115 for more details.

(e)   **Location of Service Panel** - Rule 6-206, 2-122

The service equipment must be **inside** the building served.

Fire Protection Required - Rules 6-207 & 26-444

Rule 6-207 & 26-444 require gyprock or equal fire protection behind this panel

These rules say that the service panel may not be mounted directly on wood backing. The rules also say that if the back of the panel is within 45 mm (1¾ in.) of combustible material such as wood or paper, the combustible material must be protected with a material having a flame spread rating of 25 or less.

Install a sheet of gyprock behind the panel to comply with this rule. As shown in the illustration this additional protection is required only in the area behind the panel not on any of its sides.

Rule 6-207 & 26-444 require gyprock or equal fire protection behind this panel

Service panel in shallow enclosure

Accessible - The service panel should be mounted in a free wall where it will remain accessible. It should not be located above freezers, washers, dryers, tubs, counter space, etc. It should not be on the back wall of a storage room where access may become difficult due to stored items, nor should it be in bathrooms, clothes closets, stairwells, kitchen cabinets or similar places, Rules 6-206, 2-308(1) & 26-442.

Service panel is shown recessed in an outside wall - this may not be acceptable in some cases.

Where such equipment is flush mounted in a wall, a covering door may be installed over the equipment for appearance sake. This cabinet would be very shallow, providing no storage space for other items. Sometimes a calender or large picture is hung over the panel. This always provides a certain amount of excitement when a fuse blows, all the lights go out and you have to look under calenders and pictures to find the service panel.

(f)   **Height** - Rules 6-206(1), 26-442(2)

Minimum height above floor - The code no longer specifies a minimum height for the panel. The revised rule now simply says the panel must always be placed **as high as possible** but never more than **1.7 m (67 in) to the top of the top breaker** in the panel.

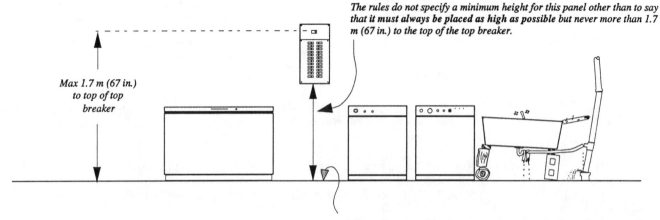

The rules do not specify a minimum height for this panel other than to say that it must always be placed as high as possible but never more than 1.7 m (67 in.) to the top of the top breaker.

Max 1.7 m (67 in.) to top of top breaker

This floor area, 1 m. (39.4 in.) in front of the panel, must be kept clear. The service panel may not be located above any appliances or counters or similar objects.

(g)   **Which end is up?**

Service circuit breaker panelboards should be mounted in a vertical position although there is no rule that actually says so, (circuit breakers will function in either the horizontal or vertical position). There is, however, the question of - - of   - well, - professionalism. Panels mounted in a vertical position do look more handsome, don't you think?.

Please note - Most fused service switches may not be mounted upside down or on their side, Rule 14-502.

(h)   **Arrangement of conductors** - Rule 6-212, 12-3036

Inside your service panel is a barrier that divides the space into two separate sections. The main service breaker is (usually) in the top section and the branch circuit breakers are (normally) in the lower section.

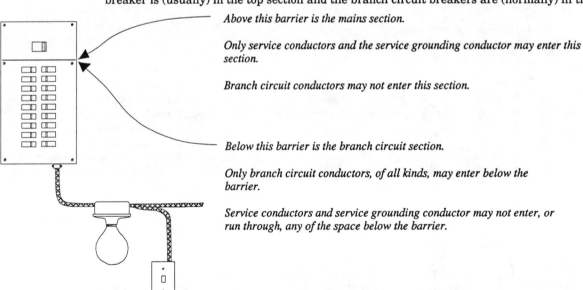

Above this barrier is the mains section.

Only service conductors and the service grounding conductor may enter this section.

Branch circuit conductors may not enter this section.

Below this barrier is the branch circuit section.

Only branch circuit conductors, of all kinds, may enter below the barrier.

Service conductors and service grounding conductor may not enter, or run through, any of the space below the barrier.

(i)   **Service Section** - may contain only the service conductors and the service ground. Branch circuit conductors may not enter, or run through, this section.

(ii)   **Branch Circuit Section** - may contain only branch circuit conductors of all kinds. Service conductors and service grounding conductors may not enter, or run through, this section.

(i)   **Grounding Electrodes** - Rule 10-002, 10-106, 10-700 & 10-702

     (i)   **The object of grounding** the electrical wiring and equipment in your house is to reduce the possibility of electrical shock and fire damage. This is a very important part of your installation. Good grounding depends on good grounding electrodes. The rules permit several different kinds to be used depending on what is available.

     (ii)   **Metal Water piping System**

     Rule 10-700 says that whenever we have at least 10 ft. (3 m) of continuously conductive metal water piping system that is located underground at least 600 mm (24 in.) below finished grade level entering a single family house we must use that metal pipe as the service grounding electrode. This arrangement is acceptable without any additional artificial grounding electrodes.

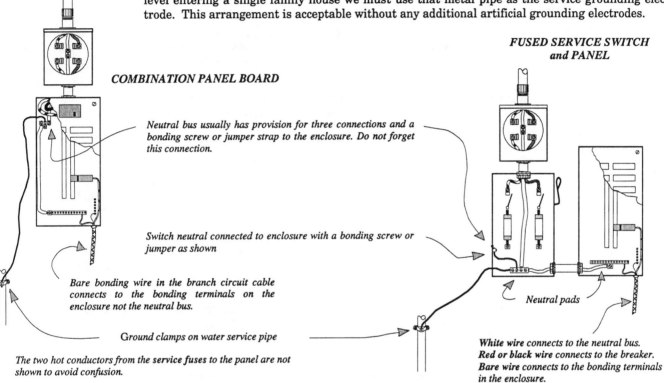

FUSED SERVICE SWITCH and PANEL

COMBINATION PANEL BOARD

Neutral bus usually has provision for three connections and a bonding screw or jumper strap to the enclosure. Do not forget this connection.

Switch neutral connected to enclosure with a bonding screw or jumper as shown

Bare bonding wire in the branch circuit cable connects to the bonding terminals on the enclosure not the neutral bus.

Ground clamps on water service pipe

The two hot conductors from the service fuses to the panel are not shown to avoid confusion.

Neutral pads

White wire connects to the neutral bus. Red or black wire connects to the breaker. Bare wire connects to the bonding terminals in the enclosure.

     (ii)   **Metal well Casing** - Where the water pump is submersible type inside a metal casing the rules permit the casing to be used as the service grounding electrode. It should be noted that the Ministry of Environment will not permit a well casing to be used as a service grounding electrode unless, as noted above, the pump motor is inside the casing. Where the pump is above ground type and is connected with nonconductive piping the well casing need not be, in fact may not be, bonded to the grounding system except as permitted by the Ministry of the Environment.

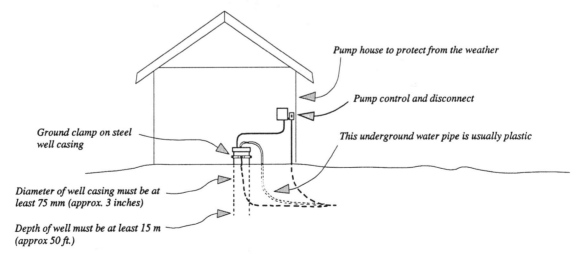

Pump house to protect from the weather

Pump control and disconnect

Ground clamp on steel well casing

This underground water pipe is usually plastic

Diameter of well casing must be at least 75 mm (approx. 3 inches)

Depth of well must be at least 15 m (approx 50 ft.)

(iii)   **Ground Rods** - Where there is no continuously conductive metal water piping system or acceptable well casing available we may use ground rods. The following picky points must be remembered when installing ground rods.

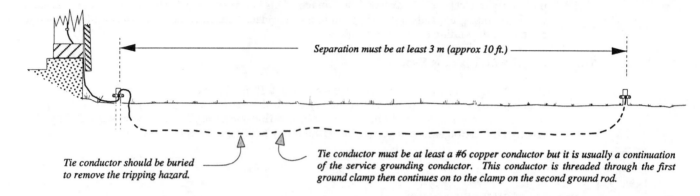

Separation must be at least 3 m (approx 10 ft.)

Tie conductor should be buried to remove the tripping hazard.

Tie conductor must be at least a #6 copper conductor but it is usually a continuation of the service grounding conductor. This conductor is threaded through the first ground clamp then continues on to the clamp on the second ground rod.

Minimum Size - $^5/_8$ inch in diameter by 3 m (10 ft.) long if of galvanized iron; or
$^1/_2$ inch in diameter by 3 m (10 ft.) long if of copper or is copper clad.

Depth - Ground rods must be driven into the ground their full length.

Spacing - Ground rods must be spaced at least 3 m (10 ft.) apart.

Tie Conductor - The rods must be connected together with a #6 or larger, copper bonding conductor. Normally, this tie conductor is a continuation of the service grounding conductor.

Ground Clamps - must be of copper or bronze if the connection is exposed to moisture or is to be buried in the ground. Dry type connectors may be used only where it will remain dry.

Tripping Hazard - In the illustration the connection to the rods is above ground. To remove this tripping hazard many inspection authorities require these connection points to be completely buried.

Rock Bottom - Can't Drive that thing into the ground? There will be locations where the ground cover is less than 3 m (10 ft.) and the ground rods can not be driven in the usual way. The rules allow the following labour intensive solutions.

Where ground cover is less than 1.2 m (48 in.) deep the rule says we must bury the ground rods at least 600 mm (24 in.) below finished grade in a horizontal trench. Where the ground cover is less than 600 mm, you, are in deep trouble. The solution is to find a location with sufficient ground cover to use one of these arrangements shown then run your grounding conductor to that location. This grounding conductor will need to be adequately buried to protect it. Where this cable must run across bare rock you will need to chip out a trench at least 150 mm (6 in.) deep. Rule 12-012(7) says the ground cable must then be laid in this trench and grouted with concrete to the level of the rock surface.

Rock bottom

The rule requires that where rock is encountered at a depth of 1.2 m (48 in.) the ground rod is to be driven to rock bottom then the remainder bent over and placed in a horizontal trench at least 600 mm (24 in.) below grade. This is more difficult but it is according to code.

This ground rod is driven into the ground at an angle. The problem with this method is not knowing what angle to drive the rod to get maximum depth. If it is driven at too steep an angle it will not be at maximum depth as required by code.

(iv)   **Concrete Encased Electrode** - consists of a single copper conductor, #4 or larger, encased in the concrete footings of the building. The copper conductor is thus in intimate contact with the concrete and thence with the earth under the footings. Properly installed these electrodes are very effective. Herein lies the key - to ensure an acceptable installation your Inspector may want to see it before he will permit it to be covered. Check with your Inspector.

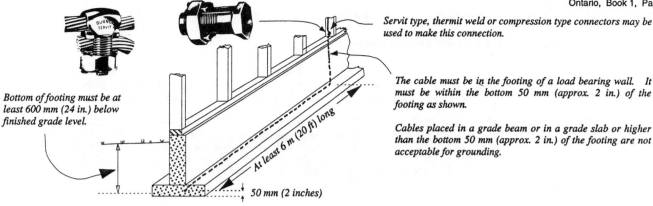

Servit type, thermit weld or compression type connectors may be used to make this connection.

The cable must be in the footing of a load bearing wall. It must be within the bottom 50 mm (approx. 2 in.) of the footing as shown.

Cables placed in a grade beam or in a grade slab or higher than the bottom 50 mm (approx. 2 in.) of the footing are not acceptable for grounding.

Bottom of footing must be at least 600 mm (24 in.) below finished grade level.

At least 6 m (20 ft) long

50 mm (2 inches)

Conductor Required

Type - must be copper.
- must be bare.

Length Must be at least 6 m (approx. 20 ft.) but may be much longer. This is the horizontal length of conductor along the base of the footing.

In addition to the 6 m length of conductor run horizontally along the bottom 50 mm (approx. 2 in.) of the footing, you will require enough length to run up to the top of the foundation wall. Allow an additional 6 in.(approx. 150 mm) or so for connection to the service grounding conductor. See illustration above.

Size- Note the following grounding conductor sizes for both the portion which is concrete encased and the part above the concrete, the home run to the panel. This cable may be spliced above the concrete foundation level as shown in the illustration above.

	Concrete encased portion	Home run to panel
Up to 125 amps	#4 copper	#6 copper
126 to 165 amps	#4 copper	#4 copper
166 to 200 amps	#3 copper	#3 copper
201 to 260 amps	#2 copper	#2 copper

Position Must be in the bottom 2 inches (50 mm) of the footing. It is not correct to place this conductor in the floor slab or under it even if it is encased in concrete. The effectiveness of this electrode depends, in part, on the weight of the building to maintain intimate contact with earth under the footing.

Connection Where the service grounding conductor must be connected to this ground electrode, such connection may be made using a split bolt type connector as shown.

(j)   **Plastic Water Service Pipe** - Rule 10-406(2) & 10-702(7)

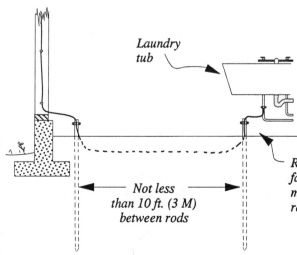

Laundry tub

Whenever the water service is with plastic pipe we cannot use it as our grounding electrode. We must in those cases install an artificial grounding electrode. Don't forget the bonding requirement. Even if the water service is plastic pipe, the piping system inside the building may be metallic. If it is, it must be bonded to the ground rods, Rule 10-406(2). This connection may be made at any convenient point on the cold water pipe where it will remain accessible.

Not less than 10 ft. (3 M) between rods

Rods project above the floor or grade level just far enough to allow connection to be made. They may also be recessed in a wall or floor if removable panels are provided for access.

(k)   **Ground Cable** - Home runs from the electrode to the service panel, Rule 10-806

#6 Insulated or Bare ground cable

This cable may be used where it is not subject to mechanical injury. Wood moulding or plastic pipe, as shown below, may be used to protect those portions which are subject to mechanical injury. This cable may be used for grounding any service up to 125 amp.

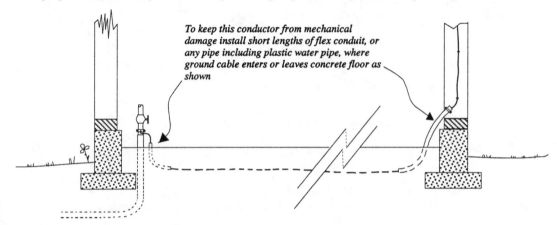

To keep this conductor from mechanical damage install short lengths of flex conduit, or any pipe including plastic water pipe, where ground cable enters or leaves concrete floor as shown

#4 Insulated or Bare ground cable

If not exposed to severe mechanical injury, may be used for grounding any size service up to 165 amp. This is the most common service ground in use today.

(l)   **Accessibility** - Rules I0-902(2) & I0-904(2)

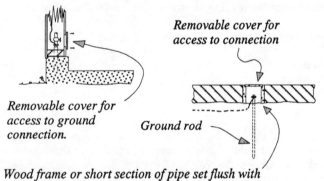

Removable cover for access to connection

Removable cover for access to ground connection.

Ground rod

Wood frame or short section of pipe set flush with surface of sidewalk or grade.

Rule 10-902(2) says that if the electrical service is grounded to the incoming water service pipe, as described above, the connection must remain accessible wherever possible. Rule 10-904(2) says the same thing about connections to ground rods. This means just what it says - leave it accessible. If the rods or the water pipe are located in a wall, provide an access panel as shown. Where the connections are a tripping hazard they may be recessed into the ground or concrete as shown.

(m)   **Ground Clamps** - Rule 10-908

Make sure that the ground clamps you install are not only CSA certified but also that they are of copper, bronze or brass. The dry type connectors will not be approved in any outdoor location.

Where the ground clamps are in a consistently dry location, such as in a dry crawl space or basement or located in a wall, as shown under "Accessibility" above, the dry type clamp may be used. It's easy to get caught on this one - watch it.

(n)   **Bonding**

(i)   **Service Switch** - Rule 10-204 — All service equipment is provided with a brass bonding screw which must he installed to connect the neutral bus in the mains section to the metal enclosure. See sketch above under "(i) Grounding Electrodes" on page 39.

(ii)   **Gas Line** - Rule 10-406(4) - Some inspection authorities consider the gas piping to the furnace, range, or dryer adequately bonded to ground with the bare bonding conductor in the supply wiring to these appliances. Other inspection authorities require a #6 copper bonding jumper between the gas piping and the service ground. Check with your local Electrical Inspector for the requirements in your area. For an illustration of one method of gas pipe bonding see page 97.

15 BRANCH CIRCUIT WIRING

(a) **Overcurrent Protection**

(i) **Light and Plug Outlets** Rules 30-104, 14-600 - The maximum rating of fuse or breakers supplying lighting or branch circuits for plug outlets is 15 amp.

(ii) **Appliance Plug Outlets** - Rule 14-600 - The maximum rating of fuses or breakers supplying appliance plug outlets in kitchens and utility rooms is 15 amp. Throw away that 20 amp fuse or breaker - you need more circuits, not bigger fuses.

(iii) **Range and Dryer Plug Outlets** - Rule 26-746 requires plug outlets for these heavy appliances. See also page 93 under "Heavy appliances".

(iv) **How Many Circuits** - Rule 30-104 - Each circuit breaker or fuse may supply only one circuit. It is not correct to connect two or more wires to a circuit breaker or fuse, Rules 6-212, 8-108, 12-3000, 26-704. A sufficient number of breaker or fuse spaces should be provided in the service panel to comply with this requirement. See also page 35 under "Minimum Circuits Required".

(v) **Fuses** - Rule 14-204 - Where fuse type panels are used it is difficult, if not impossible, to control the size of branch circuit fuses used. It is too easy to replace a blown fuse with one of a higher rating. To avoid this, the Code requires that all plug fuse holders must have a non-interchangeable adapter, as shown. Once a I5 amp adapter has been installed in a fuse base, the size of the opening and the thread is altered so that only a 15 amp fuse will now fit the base.

There is another less effective method of providing the non-interchangeable feature. This consists of a reject washer which is inserted in the fuse socket in the panel. Each washer has a different size opening which will prevent a fuse of higher rating from making contact. The weakness with these washers is that they are too easily removed. They are CSA certified and may be installed. Who knows how long they remain in the sockets.

PLUG FUSES—TYPE C

Reject Washer

(b) **Size of Cable**

Use only #14 wire unless your runs are unusually long, say more than 30 m (approx. 100 ft.) long. The #12 wire is stiff. It can cause excessive strain on the switch and receptacle terminals.

3-Wire Cable can save time and money - you are running two circuits in each cable. Rule 14-010(b) now says that we do not need tie bars or two pole circuit breakers for 3-wire cables except where these cables supply 240 volt loads, such as a range or dryer etc. or 120 volt SPLIT receptacles. This means that two single pole circuit breakers, without a tie-bar are acceptable for a 3-wire cable which supplies only lighting or plug outlets each of which is connected to the neutral and one hot conductor. **EXCEPTION** - 3-wire cables used to supply any split receptacles, such as are used on the kitchen counter, must be protected with either a two pole circuit breaker or with two single pole breakers which have their operating handles connected together with a tie-bar. This is to ensure that both circuits are de-energized for safety for anyone working on these special outlets. If you are using a fuse panel you will need a special fuse block and fuse pull for these special circuits.

For heavy appliance wiring, look under specific type.

(c) **Cables Bundled together** - Rule 4-004(10)

This rule says that where cables are "in contact with each other for distances exceeding 600 mm (24 in.) the ampacity of the conductors must be corrected by applying the factors of Table 5C."

Table 5C

1 to 3 conductors may carry	100.% load.
4 to 6 conductors may carry	80.% load.
7 to 24 conductors may carry	70.% load.
25 to 42 conductors may carry	60.% load.
43 or more conductors may carry	50.% load.

A #14 loomex cable is normally permitted to carry 15 amperes but when it is run bundled together with other cables it may have to be derated to 80% or even 70% depending on the number of conductors, (not cables) in the bunch.

Logic - Before the days of super insulated and super sealed houses cables could be bundled together without concern. There was usually sufficient air movement in the walls and the ceiling to prevent dangerous temperature rise in the bundle due to mutual heating. Today we build fully sealed sardine cans which are highly insulated against heat loss. (The BTUs your furnace produced in the fall cannot escape until you open the doors in the spring.) Under these conditions cables bundled together could overheat unless they are derated so that each conductor carries a smaller load.

The rule is aimed at cables which supply electric heat, range, dryer, water heater and any other heavy loads such as an A/C unit. The concern is not with cables supplying general residential lighting and ordinary plug outlets although the Rule could be applied to these cables as well.

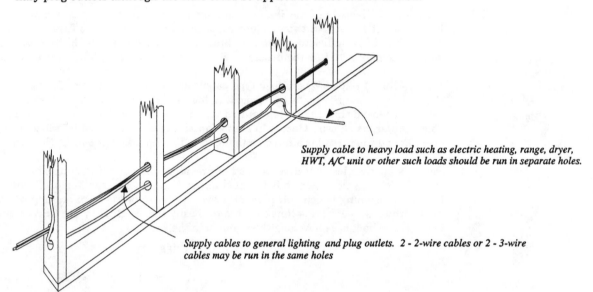

Supply cable to heavy load such as electric heating, range, dryer, HWT, A/C unit or other such loads should be run in separate holes.

Supply cables to general lighting and plug outlets. 2 - 2-wire cables or 2 - 3-wire cables may be run in the same holes

Solution

Cables supplying general lighting and plug outlets have always been derated to 80% by Rule 12-3000 and Table 5C permits 6 conductors bundled together with this 20% reduction in current carrying capacity. Therefore, cables supplying these outlets may be bundled as follows:

3 - 2-conductor cables may be run in contact without further derate, or

2 - 3-conductor cables may be run in contact without further derate

All other cables should be run in separate holes to insure separation.

Note 1   The bare bonding wire in the cables is not counted in this application.

Note 2   The derating applies only where cables are in **continuous contact** with each other for distances greater than 24 inches. Where the contact is less than 24 inches the derating does not apply. This means you could have many cables running through a hole in a stud without derating provided that the cables then fan out in different directions so that they are not in physical contact for more than 24 inches at any one point.

Caution   If you fail to observe this rule your inspector may need to ask you to restring the cables in different holes or replace them with larger cables.

The simplest solution is to force a bit of insulation or a wood chip between the cables to provide the required separation. Even though the cables are in contact with each other where they pass through holes the continuous contact in each case would then be less than 24 inches.

(d)   **Type of Cable** - Rule 30-412

All ceiling outlet boxes **on which it is intended to mount a light fixture,** (not junction boxes) must be wired with 90°C conductors such as NMD90 loomex cable. This sounds threatening, but the truth is it would be difficult to find any loomex cable intended for dry locations with a rating less than 90°C. There is one exception. NMW and NMWU cable has only a 60°C rating. NMW cable is intended for wet locations such as barns etc. NMWU cable is intended for use underground. In these damp or wet locations the rule permits these 60°C cables to enter a ceiling light outlet box which is also located in the damp or wet location..

Cold Regions - Rule 12-102, - Thermoplastic insulated cables such as NMD-90 cable can be seriously damaged if it is flexed at temperatures lower than -10 degrees C (14 degrees F). Do not install cables in cold temperatures unless the cable is specifically approved for use at that temperature. Failure to observe this requirement may result in a seriously compromised installation. It's no fun doing this work in cold weather anyhow. This work should be done in warm weather when it can be fully enjoyed.

(e)   **Cable Strapping** - Rule 12-510

Loomex cable should be strapped within 300 mm (12 in.) of the outlet box and approximately every 1.5 m (59 in.) throughout the run.

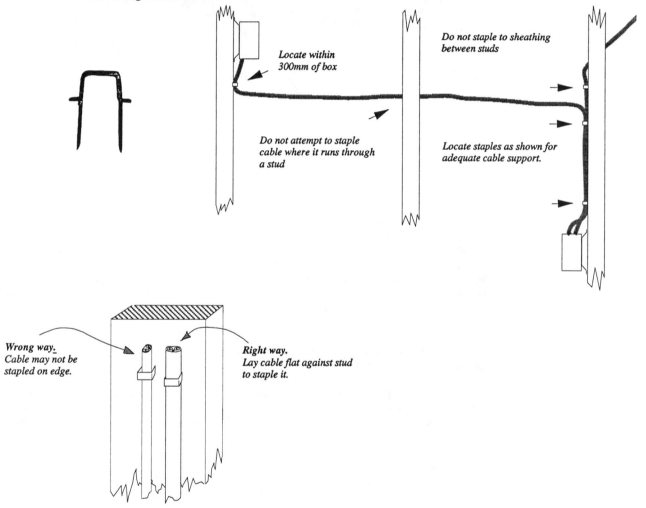

Caution

1 - **Do not staple 2 wire cable on edge** - they must lie flat. See Rule 12-506(5). This applies to two wire cables only.

2 - **Do not overdrive the staples.** Drive staples only until they contact the cable sheath - don't squish the cable. Because the cables are scantily dressed (insulated) the installers must be more careful when handling and strapping it.

3 - **Rule 2-108 - Be sure to use the correct size staple** or strap for each size cable. It is not correct to use a 2-wire cable strap on a 3-wire cable unless the staple or strap is specifically approved for both sizes, nor is it correct to put two cables under a single strap or staple. (You may get away with two cables under a strap if they are very carefully installed).

4 - Where cables are run along studs or joists they should be kept at least 1¼ in. from the nailing edge. Between staples the cable is free to move aside should a drywallers nail miss the stud but at the point of the staple the cable is held captive. If the cable has been stapled too close to the edge of a stud or joist it really needs protection. There is no code rule which specifically requires this protection except that Rule 2-108 says poor workmanship will not be accepted by the inspection department. It is best to run your cables along the middle, or as near the middle, of the stud or joist wherever possible and provide additional protection where it is not possible to run it there.

(f)   **Cable Protection** - Rule 12-516

(i)   Where the cable is run through holes in studs, plates or joists, these holes shall be at least 32 mm (1¼ in.) from the edge of the wood member.

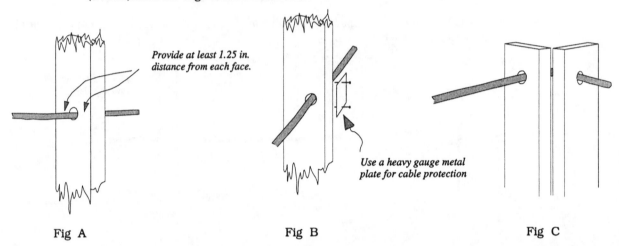

Provide at least 1.25 in. distance from each face.

Use a heavy gauge metal plate for cable protection

Fig A   Fig B   Fig C

(ii)   **In the case of small dimension members** such as may be used in partition walls, the cable hole should be located so that there is 32 mm (1¼ in.) clearance on one side. Fig.B. To protect the other side use a minimum #14 gauge metal plate (the side of a metal sectional outlet box does this job very well). This must be done in every case where the 32 mm (1¼ in.) clearance cannot otherwise be obtained. In corners of rooms, as in Fig.C., the holes may need to be drilled at an angle providing less than the minimum distance - here too, use heavy metal plates to protect the cable from drywallers nails.

Holes may contain more than one cable but must be large enough to prevent damage to the cable sheath during installation.

Caution - Bundled cables could be a serious problem. Where a number of cables are run in contact with each other you may be in conflict with Rule 4-004(10) depending somewhat on the type of loads served by those cables. See under (c) Cables Bundled together on page 43 for detailed information.

(iii)   **Attic Spaces** - Rule 12-514(a) says cables may be run on the upper faces of joists or lower faces of rafters, as shown below, provided the head clearance, joist to rafter, is 1 m (39.4 in) or less. Where the head room is greater than 1 m the cables must be run through holes

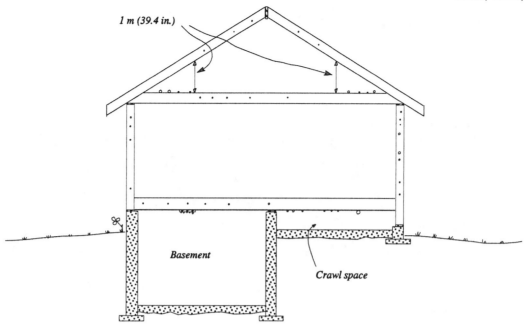

1 m (39.4 in.)

Basement

Crawl space

Run through holes in joists or studs or provide protection for the cable with a running board as shown.

(iv) **Basements and crawl spaces** - Rule 12-514(b) says cables may be run on the lower face of basement or crawl space joists provided the cables are suitably protected. A running board nailed to the underside of the joists may be an acceptable protection - check with your Inspector.

It is best to protect the cables by running them through holes drilled in the joists. This allows the basement to be finished without having to re-run cables to get them out of the way.

(iii) **Exposed Cable** - Rule 12-518 - Where loomex cable is run on the surface of the wall and within 1.5 m (59 in.) from the floor, as is often the case in buildings of solid wall construction, the cable must be protected from mechanical damage with wood or similar moulding.

(iv) **Hot Air Ducts or Hot Water Pipes** - Rule 12-506(4) requires loomex cable to be kept at least 25 mm (1 in.) away from all hot air heating ducts and hot water piping. A chunk of building insulation may be placed between the cable and the duct or pipe, as shown.

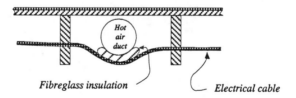

Hot air duct

Fibreglass insulation   *Electrical cable*

Ontario Bulletin 2-1-5 draws attention to two important points to watch out for. These are:

1 **Loomex cable** - (The Code calls this non-metallic sheathed cable) must not, anywhere, be in contact with metal heating ducts.

2 **Metal Heating Duct** - Where the electric wiring is installed first, before the heating ducts are in place, the cables must be carefully routed to clear the proposed route of the heating ducts. Check carefully, all cables near this duct once both systems are installed. Rule 12-010(5) says it's okay to run in, or just through, the cold air return if this consists of an **unlined joist space** but, again, the cable must not contact the metal sheathing nailed to the underside of the joists to make the duct.

16   CHART of BRANCH CIRCUITS

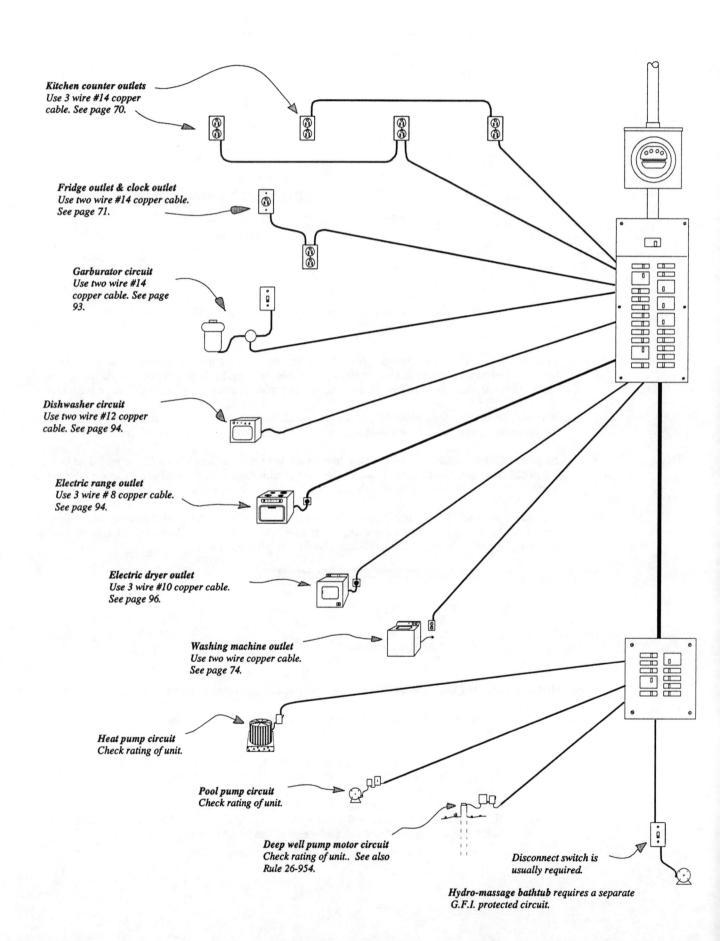

Kitchen counter outlets
Use 3 wire #14 copper cable. See page 70.

Fridge outlet & clock outlet
Use two wire #14 copper cable. See page 71.

Garburator circuit
Use two wire #14 copper cable. See page 93.

Dishwasher circuit
Use two wire #12 copper cable. See page 94.

Electric range outlet
Use 3 wire # 8 copper cable. See page 94.

Electric dryer outlet
Use 3 wire #10 copper cable. See page 96.

Washing machine outlet
Use two wire copper cable. See page 74.

Heat pump circuit
Check rating of unit.

Pool pump circuit
Check rating of unit.

Deep well pump motor circuit
Check rating of unit.. See also Rule 26-954.

Disconnect switch is usually required.

Hydro-massage bathtub requires a separate G.F.I. protected circuit.

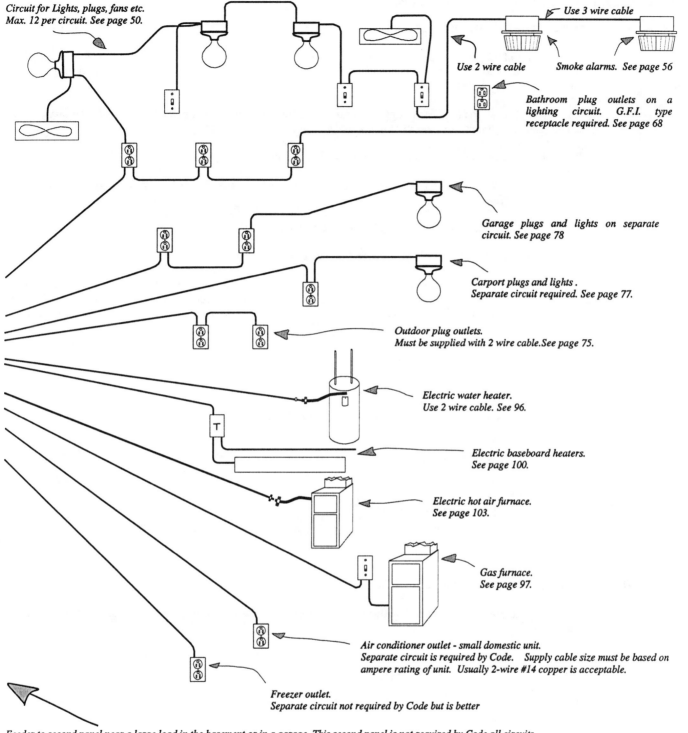

Circuit for Lights, plugs, fans etc.
Max. 12 per circuit. See page 50.

Use 3 wire cable

Use 2 wire cable

Smoke alarms. See page 56

Bathroom plug outlets on a lighting circuit. G.F.I. type receptacle required. See page 68

Garage plugs and lights on separate circuit. See page 78

Carport plugs and lights .
Separate circuit required. See page 77.

Outdoor plug outlets.
Must be supplied with 2 wire cable.See page 75.

Electric water heater.
Use 2 wire cable. See 96.

Electric baseboard heaters.
See page 100.

Electric hot air furnace.
See page 103.

Gas furnace.
See page 97.

Air conditioner outlet - small domestic unit.
Separate circuit is required by Code.   Supply cable size must be based on ampere rating of unit. Usually 2-wire #14 copper is acceptable.

Freezer outlet.
Separate circuit not required by Code but is better

Feeder to second panel near a large load in the basement or in a garage. This second panel is not required by Code all circuits can be supplied from one main panel. See page 36

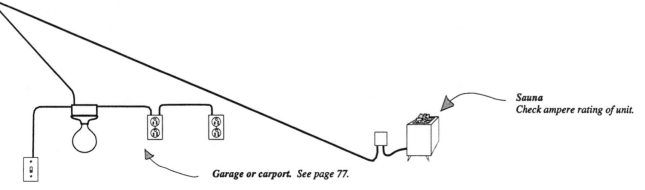

Sauna
Check ampere rating of unit.

Garage or carport. See page 77.

17. OUTLET BOXES - for Lights, Smoke Alarm & Fans

(a)   **Outlets Per Circuit - Rule 12-3000**

A maximum of 12 outlets may be connected to a circuit. This may consist of 12 light outlets or 12 plug outlets (not appliance plugs, see page 70 for appliance plugs) or any combination of light and plug outlets mixed, as long as their total number does not exceed 12 outlets.

It is better to have the load consist of a mixture of lights and plugs. This gives better load diversity on the circuit and less chance of a complete blackout in case of circuit failure.

To avoid confusion and costly duplication proceed as follows:

(i)   Make a floor plan of your house. If more than one floor, draw a separate plan for each floor.

(ii)   Show each outlet using the symbols given below.

(iii)   Determine the best location for service equipment.

(iv)   Draw a line showing the course each cable run will take. Start with one circuit and complete it before going on to the next. Identify each circuit and each outlet for quick easy location. For example, the outlets on the first circuit would be A 1, A2, A3 etc. The outlets on the next circuit would be B1, B2, B3 etc.

Symbols usually used are:

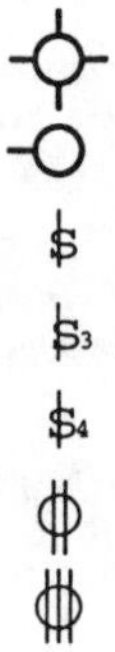

Symbol	Description
⌀	*Light outlet on ceiling*
⊸○	*Wall light outlet such as over a bathroom vanity*
₴	*Wall switch*
₴₃	*Three way switch*
₴₄	*Four way switch*
⬦	*Duplex plug receptacle*
⬦	*Split duplex receptacle used above kitchen counter see page 70*

(b)   **Light Outlet Boxes** - Rule 2-314 & 30-328

The Electrical Code is concerned with lighting outlets and their control in only two areas in the house. These are:

(i)   **The Service Panel** - Rule 2-314 requires adequate illumination at service equipment. Make sure you have a light outlet near the service panel for maintenance purposes.

(ii)   **Stairway Lighting** - Rule 30-328. All stairways must have illumination - there is no exception to this Rule.

Stairway lighting Control Switches - Rule 30-328 & Ontario Building Code Rule 9.34.2.3. These rules say that it must be possible to control stairway lights from both the head and foot of the stairway. There is an exception. Where the stairway has only three or fewer risers or where the basement is a dungeon like space which has no finished area and no egress other then the one stairway leading into it one switch at the head of the stairway is acceptable.

Watch this one! The Rule says you must be able to turn off that stairway light from either the top or bottom of the stairway. In the case of an open stairway leading to a living room, or other such area on a different level, the normal switch locations for general lighting in that area may not also meet the requirements for stairway lighting. Additional switching may need to be installed or separate lighting and control may need to be installed for the stairway. Note that this special switching requirement applies only in cases where the stairway consists of four or more steps.

(ii)   **The Ontario Building Code** - requires lighting outlets as follows:

Rule 9.34.2.1. requires a fixed lighting outlet at each doorway leading in or out (depending on which you are going). This light outlet must be controlled from a fixed wall switch located inside the house.

Rule 9.34.2.2. requires the following outlets:

Kitchen - requires at least one light outlet complete with wall switch.

Bedrooms & Living Rooms - each must have at least one light outlet, with fixture, or one plug outlet - you have a choice. This outlet must be wall switched. If you choose to install a light outlet it must be wall switched. If you also install plug outlets, and these are required by the Electrical Code rules, you would then not need to control any of the plug outlets in that bedroom or living room with a wall switch. But if you have no light outlet in a bed room or living room, (and that is still a choice) one or more of the plug outlets would then need to be wall switched to comply with this rule.

Utility rooms, Laundry rooms, Dining rooms, Bathrooms, Vestibules and Hallways. A light outlet with fixture is required in each of these rooms and each light fixture is required to be wall switched.

Stairway lighting - The requirements of Ontario Building Code Rule 9.34.2.3. is the same as Rule 30-328 in the Electrical Code. See above for clarification.

Unfinished Basement Lighting - Building Code Rule 9.34.2.4. says that "A lighting outlet with fixture shall be provided for each 30 m² (323 sq.ft.) or fraction thereof of floor area in unfinished basements". Note that the rule does not say there must be a lighting fixture "in" each 30 m² (323 sq.ft.) in the basement. Compute the total floor area of the basement then divide by 323 to determine the minimum number of lighting fixture required by this rule. These light outlets must be spaced to provide reasonably uniform light distribution over the whole floor area.

Note - The light outlet nearest the stairway must be **wall switched** at the **head of the stairs.**

Storage rooms - Rule 9.34.2.5. This rule requires a lighting outlet with fixture in each storage room. The rule does not require this light outlet to be wall switched but that is best.

Garages and Carports - Rule 9.34.2.6. This rule requires a light outlet with fixture in each garage and carport. If the fixture is mounted above the car parking space it must be wall switched. If it mounted off to the side it may be switched lampholder type.

Note - In the case where other entranceway lighting also provides light for the carport area the rule says we do not need to install any additional lighting in the carport.

Simple Light Outlet

The standard light outlet consists of a loomex cable run into an octagon or round outlet box set flush with the ceiling finish.

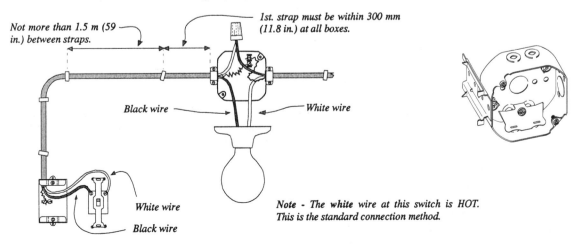

Not more than 1.5 m (59 in.) between straps.

1st. strap must be within 300 mm (11.8 in.) at all boxes.

Black wire

White wire

White wire

Black wire

Note - The white wire at this switch is HOT. This is the standard connection method.

(c) **Recessed Light Outlet** - Rules 30-308, 30-410 & Ontario Building Code Rule 9.34.1.4

NON - Insulated areas

Watch this one carefully. If there is any possibility of building insulation being placed near this fixture the Inspector will reject it unless the fixture is specifically certified and is so marked to show it is acceptable for blanketing with building insulation. Ceilings which have a living floor area above, such as a basement or first floor ceiling in a two story house, are usually not insulated. In these cases a standard CSA certified recessed light fixture may be used.

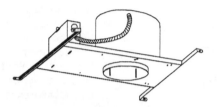

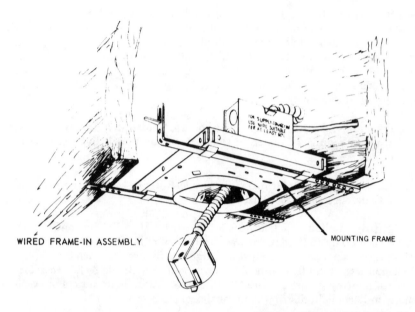

WIRED FRAME-IN ASSEMBLY

MOUNTING FRAME

Pre-Wired Type - Most recessed light fixtures used today are pre-wired type. This fixture has a short length of high temperature wire connecting the fixture socket to its connection box. The supply cable may run directly into the connection box as shown below.

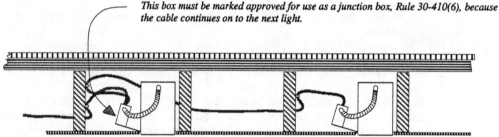

This box must be marked approved for use as a junction box, Rule 30-410(6), because the cable continues on to the next light.

The illustration shows a second floor above the recessed light fixture. In this case there is no requirement for building insulation and therefore, a standard recessed light fixture may be used. However, this fixture can still be a fire hazard if the minimum ½ in. clearance from wood is not maintained all around and on top of the fixture. The only points where the ½ in. clearance is not required is where the ceiling finish material butts up against the fixture and at the support points around the lower edge.

INSULATED Ceilings - USE SPECIALLY DESIGNED LIGHT FIXTURES

If you plan to install recessed light fixtures in an **insulated ceiling** you must use a fixture which is certified for **blanketing with building insulation.**

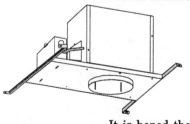

This is a specially designed fixture which is enclosed in a metal box. It may be covered with building insulation.

Lamp Wattage Rating - At present fixtures certified for covering with building insulation are limited to 75 watt lamps. In some cases a simple change in fixture trim will increase permitted lamp rating - check with your supplier.

It is hoped the industry will soon have higher wattage fixtures which are fully certified for blanketing with building insulation.

This fixture is equipped with a built-in thermostat to shut itself off in case it overheats. Some homeowners have installed 150 watt lamps in these 75 watt fixtures. It works fine, for a few minutes until the higher wattage lamp overheats the fixture. At this point the thermostat shuts it off. When it cools it turns itself back on again. Obviously, it could not do this for very long. The thermostat is not designed for repeated on off switching, it would very soon fail. Make sure you use the correct size lamp in your fixture.

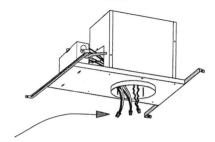

All connection leads must be long enough to reach drown through the fixture opening for future maintenance purposes.

Connection Leads - Pay particular attention to the length of the connection leads on this fixture - they are long and for a very good reason, DO NOT CUT THEM. Note too, the message on the connection box cover. It will tell you to "leave at least 12 inches of leads", or similar words. This refers to the supply cable leads you will install. These must be as long as the long fixture leads. The reason for the unusual length of these leads is to allow access to the splices after the fixture has been installed. Once the ceiling finish is in place access may be difficult, if not impossible, from above, (from the attic space). It must be possible to reach through the fixture opening, from below the ceiling level, and remove the snap-on cover plate from the connection box. It must then be possible to pull the splices out through the fixture opening and down to a point below the ceiling finish for maintenance purposes. This is why both the supply cable leads and the fixture leads provided by the manufacturer must be at least 12 inches long.

Connection or Terminal Box - The connection box on a recessed light fixture may be used only for the supply conductors to this one light fixture. You may not use it as a junction box for any other loads unless the fixture connection box is approved for that purpose. You will find many of these fixtures are equipped with boxes which are certified for use as a junction box. If they are acceptable for use as a junction box they will be so marked.

STANDARD RECESSED LIGHT FIXTURES

Fire Hazard - Improperly Installed Recessed Light Fixtures

A number of fires have been caused by incorrectly installed recessed light fixtures. There are two problem areas to watch out for when installing recessed light fixtures. First, make sure you have provided the required minimum ½ inch clearance from combustible surfaces all around the fixture. Second, make sure all recessed light fixtures installed in insulated areas are properly certified for blanketing with building insulation. Look for the manufacturers label which clearly states the fixture is acceptable for blanketing with building insulation. Without this label the fixture should be rejected for that location. The Inspector is working to make your home safe. Because these fixtures can be a fire hazard you really want him to be very careful.

The insulation does exactly what it is supposed to do - it traps the heat in the fixture. When the fixture reaches the combustion temperature of the wood or paper next to it the result is charring and sometimes fire. Use only fixtures certified and marked approved for blanketing with building insulation.

Combination Heat Lamp and Fan Fixture - CSA has advised that this combination fixture may be blanketed with building insulation.

Personal Opinion - If these fixtures are not CSA certified for direct covering with building insulation it may be dangerous to cover them. This particular fixture is usually equipped with a 250 watt heat lamp which generates a great deal of heat. As long as the fan continues to operate it will tend to keep the internal temperature of the fixture to a safe level. In the event the fan fails to operate, for whatever reason, fixture temperature may rise above the safe level. The ducting from this fan to outdoors will provide some natural ventilation provided there is no automatically operated flapper valve in this duct which is closed when the fan is not operating. It would also be necessary for the ducting to be adequately inclined to provide a chimney effect for heated air to move away from the fixture. Finally the Building Code requirements for very well sealed houses may stifle air movement to a point where you would have a 250 watt heater in the ceiling covered with R40 insulation. If the fan fails .... who knows how long it would take start a fire.

(d)   **Fluorescent Light Outlets** - Rules 12-506(1), 12-3002(5), 30-312(3)

Loomex cable may be run directly into a fluorescent light fixture as shown. Where fluorescent fixtures are mounted end to end as in valance or cornice lighting, the loomex cable should enter only the first fixture. The interconnecting wires between fixtures must be R90 or better.

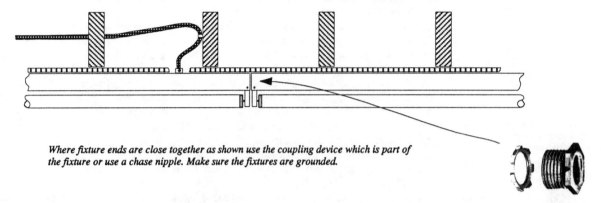

Where fixture ends are close together as shown use the coupling device which is part of the fixture or use a chase nipple. Make sure the fixtures are grounded.

That threaded thing with a locknut next to it, in the illustration, is called a chase bushing. Punch out the knockout holes in the end of the fixtures where they join. The fixtures are then fastened to the ceiling with their ends as close together as possible. The chase nipple is now inserted through the two knockouts and the locknut is used to bring the two fixture end plates together for grounding. This chase bushing also provides a smooth throat to protect the fixture supply conductors.

Outlet Box Not Required

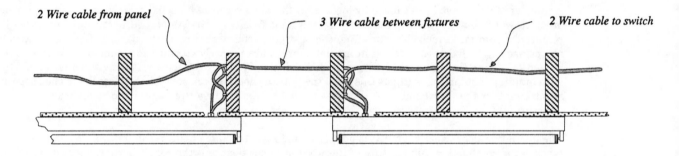

2 Wire cable from panel

3 Wire cable between fixtures

2 Wire cable to switch

As shown, the loomex cable is run directly into the fixture. An outlet box is not required provided the cable used is NMD7 or NMD90 and not more than two cables enter any fixture.

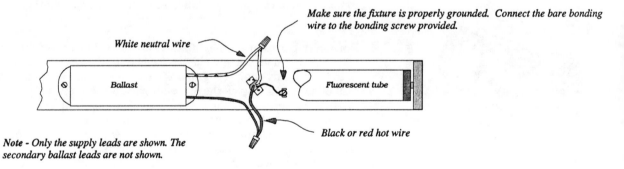

Make sure the fixture is properly grounded. Connect the bare bonding wire to the bonding screw provided.

White neutral wire

Ballast

Fluorescent tube

Black or red hot wire

Note - Only the supply leads are shown. The secondary ballast leads are not shown.

(e)   **Bathroom Light Outlets** - Rule 62-110(1)(b)

Heat Lamps - Like any other recessed light fixture, heat lamps can be a very real fire hazard if improperly installed. Care should be taken to:

(i) **Locate the Heat Lamp fixture** away from the door so that it cannot radiate heat directly onto the upper edge of the door when it is in the open position. This applies to the shower stall doors or curtain rod as well as the bathroom entry door. The rule does not specify a distance but some Inspection Authorities require at least 12 inches, horizontal measurement, between the edge of the fixture and a shower rod or a door in any position.

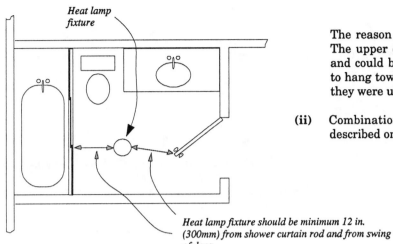

Heat lamp fixture

Heat lamp fixture should be minimum 12 in. (300mm) from shower curtain rod and from swing of door.

The reason for all this is to eliminate a possible fire hazard. The upper edge of the door would be too close to the fixture and could become overheated. The shower rod could be used to hang towels and clothing. They could become overheated if they were under the heat lamp.

(ii) Combination heat lamp/fan fixture for use in a bathroom is described on page 53.

Swag Lamps in a Bathroom - Rule 10-514(2) - Be sure to use the correct fixture - it must have a ground conductor to each chain hung lamp holder.

(f) **Clothes Closet Light Outlets** - Rule 30-204(l)

Light outlet boxes in closets may be in the ceiling or on the front wall above the door.

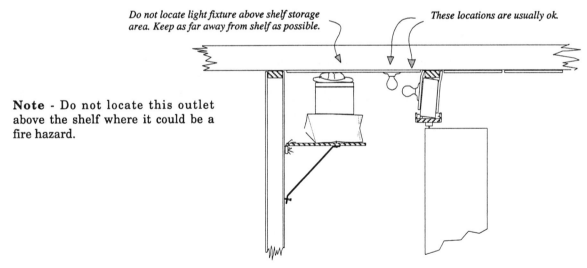

Do not locate light fixture above shelf storage area. Keep as far away from shelf as possible.

These locations are usually ok.

Note - Do not locate this outlet above the shelf where it could be a fire hazard.

(g) **Smoke Alarms** - Rule 32-110

The Ontario Building Code requires smoke alarm devices in each residential dwelling unit.

(i) **How Many Required** - The rule says they "shall be installed between each sleeping area and the remainder of the dwelling unit; and where the sleeping areas are served by hallways, the smoke alarms shall be installed in the hallway". All the bedrooms facing onto a common hallway could be served with one device, however, bedrooms on another floor or in the basement are in another "area" and would therefore require another device. The number of smoke alarm units required for your house, and the location of these devices is controlled by the local Building Inspector or Fire Officer.

(ii)   **Outlet Box Required** - Use a standard light outlet box mounted as for a light outlet. This box may also be used as a junction box to serve other loads, it need not be at the end of a run.

(iii)   **Position** - The rule says the smoke alarm shall be installed "on or near the ceiling" depending on the installation instructions that come with the device.

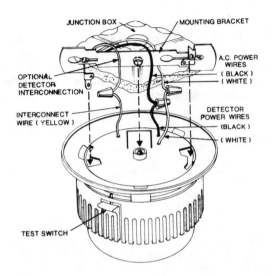

Circuit Required - May be supplied with any general lighting circuit or plug outlet circuit (except those circuits which are G.F.I. protected or which supply kitchen plug outlets, laundry or outdoor plug outlets).

Note - Do not switch this outlet. It must not be possible to turn this thing off except with the breaker in the panel.

(iv)   **2 or More Smoke Alarms** - Where two or more smoke alarms are installed the Building Code says they must all operate together - that is, if one alarm is activated to sound an alarm the others must all be connected together so that they all automatically sound the alarm together.

Units designed for line voltage (120 volts) require a 2-wire #14 supply cable to the first unit, then 3-wire from there to all the other units. The third conductor in these cables is for the signal circuit so that all units can sound the alarm together, the other two conductors are required to supply power to the second and third units. This type is in common use today.

(h)   **Overhead Fans**

These fans are becoming more and more popular. They not only look smart they also serve a useful purpose by moving the heated air downward to the floor level.

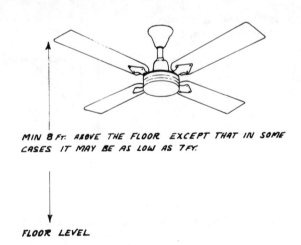

MIN 8 FT. ABOVE THE FLOOR EXCEPT THAT IN SOME CASES IT MAY BE AS LOW AS 7 FT.

FLOOR LEVEL

Some things to watch for:

(1)   **Look for a** certification label. Never purchase any electrical appliance unless it is clearly marked with a CSA or one of the other certification labels described on page 2. This is your protection and assurance that the device has been checked against a good standard.

(2)   **Look for mounting instructions.** - Each fan has a caution marking which gives the minimum mounting height above floor required for that particular fan. This marking will look something like this;

         Caution: Mount with the lowest mounting parts at least 8 ft. above floor or grade level.

Some fans will not turn as fast at maximum speed or the fan blades are designed so they are less hazardous to anyone coming in accidental contact with the blades. These fans will also have caution markings similar to the words given above except that in this case the minimum mounting height will be 7 ft. instead of 8 ft. In this case the fan blades may be as low as 7 ft. above the floor.

Near Stairways, Balconies and such like.

The fan blades may not be within reach of a person standing on a stairway, a landing or a balcony. If the blades are within reach the minimum height given on the caution notice must be measured from the level the person is standing on. Check this detail carefully in the rough wiring stage - it is very difficult to change the location of the supply outlet later.

Circuit required

This fan may be supplied with any lighting circuit which has only 11 or fewer outlets. The fan outlet, though it is a small load, counts as one outlet.

     Caution;   CSA has issued a caution regarding these fans. It warned that if the blades are not properly installed they may work loose and fall to the floor. Anyone in the path of such a flying blade could be seriously hurt.

18 Switch Outlet Boxes

(a) **Height of Switches** - The rules do not specify a required height for wall mounted light switches. They may be located at any convenient height - usually they are set at approximately 1.2 m (approx. 48 in.) to the lower edge of the box.

(b) **Bathroom Light, Heat Lamp and Fan Switches** - Rule 30-326(3) says the light switch must not be located within reach of a person in a shower or bathtub. Appendix B for Rule 30-326(3), page 604 in the Code, defines out of reach as 39.3in. (1m). This means that switches controlling these loads may be in the bathroom provided they are at least 39.3in. from the nearest inside face of a bathtub or shower stall.

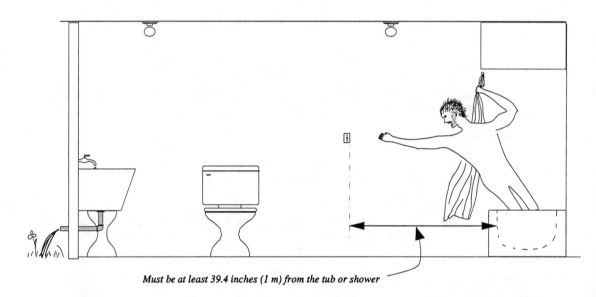

Must be at least 39.4 inches (1 m) from the tub or shower

Note 1 - This is a horizontal measurement from the switch to the nearest inside face of a tub or shower stall.

2 - **The rule does not** specifically say that this applies to all switches in a bathroom. All switches and thermostats located in a bathroom are equally dangerous if they operate at full line voltage.

(c) **Stairway Lighting Control Switches** - Rule 30-328 requires illumination at all stairwells - there is no exception. Control switches are required at both head and foot of each stairway in all cases except where the basement is a dungeon like space which has no finished area and no egress other than the one stairway leading into it.

Watch this one! The Rule says you must be able to turn off that stairway light from either the top or bottom of the stairway. This may not be possible with area lighting in the case of open stairways leading into living rooms or other such areas. See page 50 & 51 for more details.

(d) **Connection of Switches** - Rules 4-034(2), 30-602 - Switch connections shall be made so that there is a white wire and a black wire to the fixture. To do this the connection should be made as follows: The white wire in the supply cable shall connect directly to the screw shell (the silver terminal) in the lamp holder. The black wire from the switch connects to the center (gold) terminal in the lamp holder. In this way the fixture has a black and a white wire supplying it. It also has the black wire connected so that the screw base of the bulb cannot become energized. This is very important,

The following drawings illustrate acceptable methods of connection in switch and light outlet boxes.

Simplest Switching Arrangement — Power entering light outlet box first.

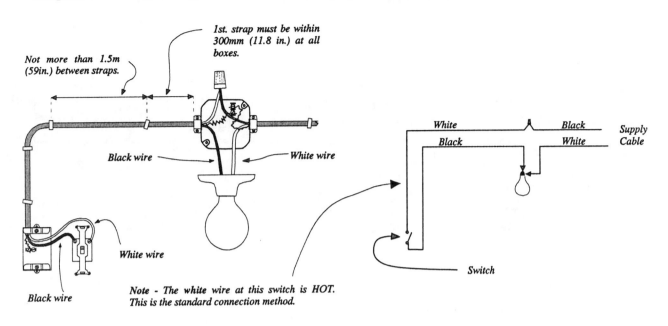

1st. strap must be within 300mm (11.8 in.) at all boxes.

Not more than 1.5m (59in.) between straps.

Black wire   *White wire*

White wire

Black wire

Note - The white wire at this switch is HOT. This is the standard connection method.

White   *Black*   *Supply Cable*

Black   *White*

Switch

Power Entering Switch Outlet Box First

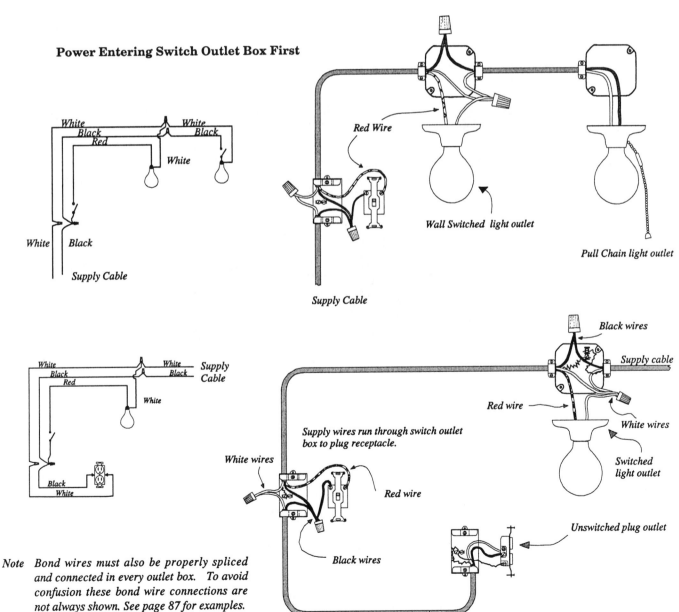

White   *White*

Black   *Black*

Red

White

White   *Black*

Supply Cable

Red Wire

Wall Switched light outlet

Pull Chain light outlet

Supply Cable

White   *White*   *Supply Cable*

Black   *Black*

Red

White

Black

White

Supply wires run through switch outlet box to plug receptacle.

White wires

Red wire

Black wires

Black wires   *Supply cable*

Red wire

White wires

Switched light outlet

Unswitched plug outlet

Note *Bond wires must also be properly spliced and connected in every outlet box. To avoid confusion these bond wire connections are not always shown. See page 87 for examples.*

3-Way Switches - Both switches on the same side of the light.

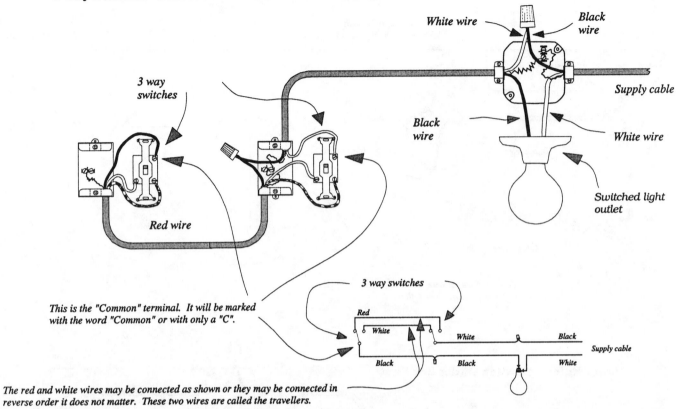

This is the "Common" terminal. It will be marked with the word "Common" or with only a "C".

The red and white wires may be connected as shown or they may be connected in reverse order it does not matter. These two wires are called the travellers.

Note Bond wires must also be properly spliced and connected in every outlet box. To avoid confusion these bond wire connections are not always shown. See page 87 for examples.

3-Way Switches - Light outlet between switches

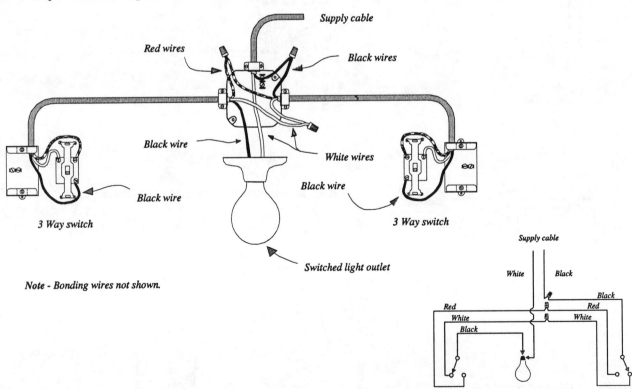

Note - Bonding wires not shown.

4-Way Switch Control - for two light outlets.

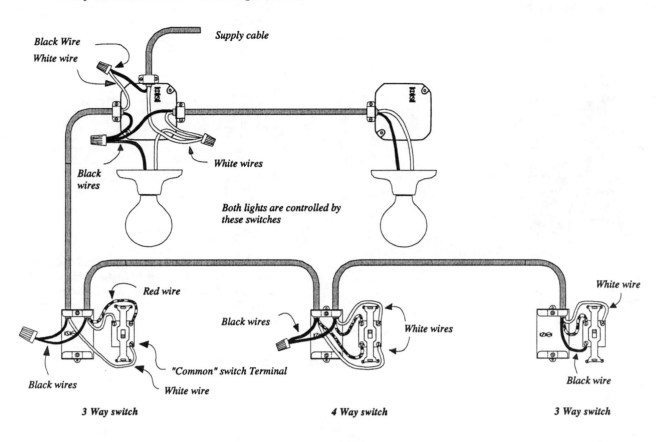

Supply cable

Black Wire
White wire

White wires

Black wires

Both lights are controlled by these switches

Red wire

Black wires

White wires

White wire

"Common" switch Terminal

Black wires

White wire

Black wire

3 Way switch

4 Way switch

3 Way switch

Note - Ground wires are not shown to avoid confusion

There are two types of 4-way switches. The crossed wire type and the through wire type. The illustration shows the through wire type which is the most common type in use today.

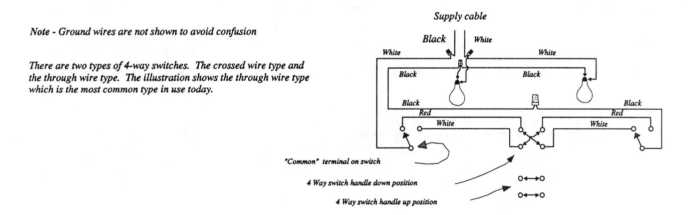

Supply cable

Black   *White*

White   *White*

Black   *Black*

Black   *Black*

Red   *Red*

White   *White*

"Common" terminal on switch

4 Way switch handle down position

4 Way switch handle up position

Note   Bond wires must also be properly spliced and connected in every outlet box.   To avoid confusion these bond wire connections are not always shown. See page 87 for examples.

2-Gang Switch Box

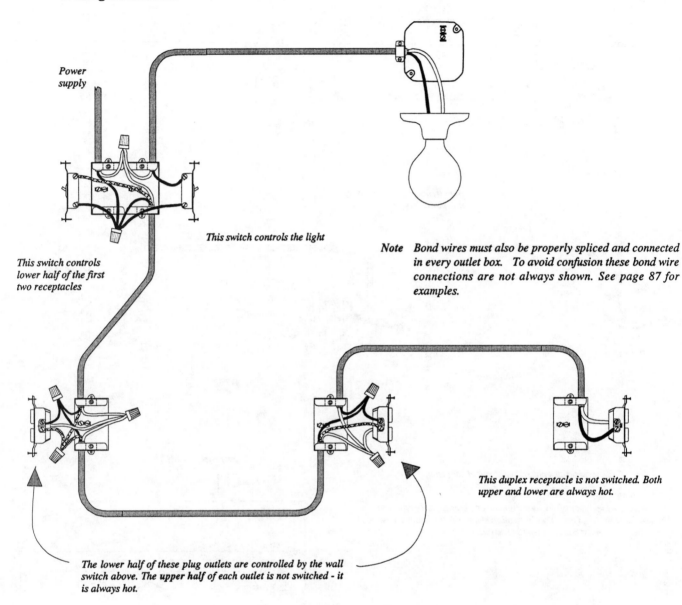

Power supply

This switch controls the light

This switch controls lower half of the first two receptacles

Note Bond wires must also be properly spliced and connected in every outlet box.   To avoid confusion these bond wire connections are not always shown. See page 87 for examples.

This duplex receptacle is not switched. Both upper and lower are always hot.

The lower half of these plug outlets are controlled by the wall switch above. The upper half of each outlet is not switched - it is always hot.

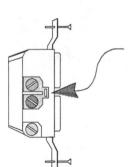

Note - the small break away tab on these two receptacles must be removed. If this is not done the above switch will not work and both halves of the duplex receptacle will always be hot at the same time.

Note - The bare bond wires are not shown to avoid confusion.

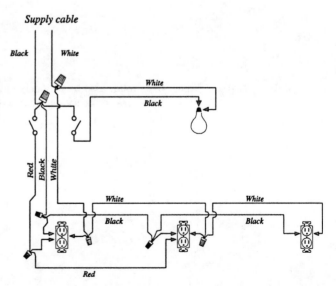

Supply cable

Black   *White*

White

Black

Red   *Black*   *White*

White   *White*

Black   *Black*

Red

Some typical branch circuits.

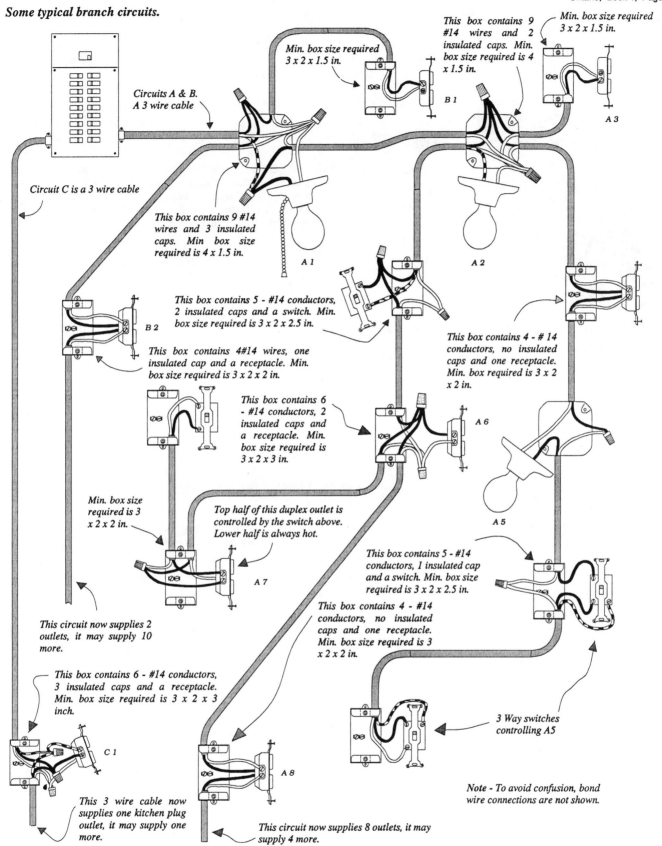

Min. box size required 3 x 2 x 1.5 in.

This box contains 9 #14 wires and 2 insulated caps. Min. box size required is 4 x 1.5 in.

Min. box size required 3 x 2 x 1.5 in.

Circuits A & B. A 3 wire cable

B 1

A 3

Circuit C is a 3 wire cable

This box contains 9 #14 wires and 3 insulated caps. Min box size required is 4 x 1.5 in.

A 1

A 2

This box contains 5 - #14 conductors, 2 insulated caps and a switch. Min. box size required is 3 x 2 x 2.5 in.

B 2

This box contains 4 - #14 conductors, no insulated caps and one receptacle. Min. box required is 3 x 2 x 2 in.

This box contains 4#14 wires, one insulated cap and a receptacle. Min. box size required is 3 x 2 x 2 in.

This box contains 6 - #14 conductors, 2 insulated caps and a receptacle. Min. box size required is 3 x 2 x 3 in.

A 6

A 5

Min. box size required is 3 x 2 x 2 in.

Top half of this duplex outlet is controlled by the switch above. Lower half is always hot.

A 7

This box contains 5 - #14 conductors, 1 insulated cap and a switch. Min. box size required is 3 x 2 x 2.5 in.

This box contains 4 - #14 conductors, no insulated caps and one receptacle. Min. box size required is 3 x 2 x 2 in.

This circuit now supplies 2 outlets, it may supply 10 more.

This box contains 6 - #14 conductors, 3 insulated caps and a receptacle. Min. box size required is 3 x 2 x 3 inch.

C 1

A 8

3 Way switches controlling A5

This 3 wire cable now supplies one kitchen plug outlet, it may supply one more.

This circuit now supplies 8 outlets, it may supply 4 more.

Note - To avoid confusion, bond wire connections are not shown.

Note Circuits A & B are supplied with a 3-wire cable to the first outlet where it splits into 2-wire cables. Each of the outlets, supplied by these cables, is connected to one hot wire and the neutral, therefore, the circuit breakers supplying this 3-wire cable do not need to be equipped with a tie-bar. The 3-wire cable on the left supplies kitchen counter outlets. Each of these is connected to both hot wires and the neutral, therefore, this cable must be supplied with either a two pole circuit breaker or two single pole breakers with their operating handles tied together with a tie-bar.

19 Plug Outlets - Rule 26-702

(a) **Height** - ,the rules do not specify any definite height for plug outlets. They may be at any convenient height. Usually they are placed at approximately 300 mm (approx. 12 in.) to lower edge of the outlet box in the living room, dining room, bedrooms and hallways etc.

Horizontal or Vertical - may be either way but if you want a professional looking job, install them all in the vertical plane.

(b) **How Many Plugs and Where Required**

Living room
Family room
Rec. room
Bedroom
Den
Study

Rule 26-702(4)&(5) require plug outlets in these rooms to be located so that it is not possible for an electrical appliance to be more than 1.8 m (approx. 6 ft.) from a plug outlet when it is located anywhere along the wall.

Note - This measurment is not a radius - you must measure into corners as shown in the illustration. This is the strict interpretation of Subrule 2.

Notes Wall space less than 900 mm (approx. 36 in.) wide is not required to have an outlet.

Rule 26-702(5) says do not count spaces occupied by:

(1) Doorways, and the area occupied by the door when fully open.

(2) Window - The space occupied by windows that extend to the floor need not be counted.

(3) Fireplaces.

(4) Other permanently fixed installations which limit the use of wall space.

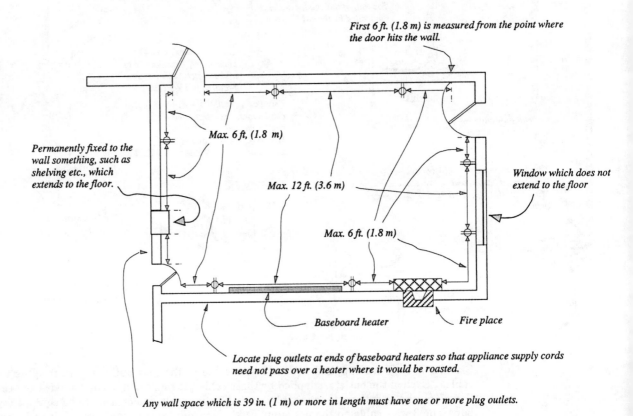

First 6 ft. (1.8 m) is measured from the point where the door hits the wall.

Permanently fixed to the wall something, such as shelving etc., which extends to the floor.

Max. 6 ft, (1.8 m)

Max. 12 ft. (3.6 m)

Max. 6 ft. (1.8 m)

Window which does not extend to the floor

Baseboard heater

Fire place

Locate plug outlets at ends of baseboard heaters so that appliance supply cords need not pass over a heater where it would be roasted.

Any wall space which is 39 in. (1 m) or more in length must have one or more plug outlets.

Electric & Other Types of Baseboard Heaters - are a permanent unit and they do limit the use of that wall space to some extent, however, appliances such as radios, T.V., swag lamps, etc. can be placed in that wall space and each of these requires power. The rule therefore, does require outlets in wall spaces occupied by baseboard heaters. These outlets should not be located above the heaters unless it cannot be avoided. Usually they can be located at the ends of the heaters so that electrical supply cords need not run over the heater and be roasted every time the heater comes on. See also Appendix B on page 599 in the Electrical Code for confirmation of this recommendation.

(c)   **Entrance (Foyer)** - Rule 26-702(2) - If this is a room, treat it as a living room. If it is like a hallway, apply the rule for hallways. If it is something in between - well - just put in the extra outlets and be done with it - don't quibble over little things.

(d)   **Hallways**- Rule 26-702(9) - Locate the plug outlet so that no point on the hallway floor is more than 4.5 m (15 ft) from a plug outlet without having to to go through a doorway fitted with a door.

   Note   A short open type hallway (such as between a living room and kitchen where there are no doors) does not require a plug outlet at all, provided that no point in the hallway is more than 4.5 m (approx. 15 ft.) from a plug outlet in either of the rooms at the ends of the hallway.

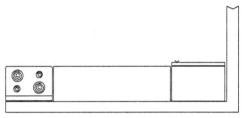

Short open hallway. No point in the hallway is more than 4.5 m (177 inches) from an outlet.

(e)   **Basement Wiring** - Rule 26-702

   Finished Basement - The rules require the same number of outlets along the walls of a basement as for similar rooms upstairs. The walls of basement bedrooms, hallways, family rooms etc. which are finished must be wired as similar rooms on the main floor.

   Unfinished Basement

   First A Definition of Unfinished - Rule 26-702(1)

   This Rule defines an unfinished basement as follows:

   (1)   **If the wall finish material** does not extend fully to the floor, ie. if the lower 450 mm (17.7 in.) of the wall is not finished with any kind of finishing material that wall is considered to be unfinished. Such a wall is required to have only minimum wiring as described below. Building insulation and vapour barrier may be installed in all walls and it may extend to the floor. Building insulation is not finishing material.

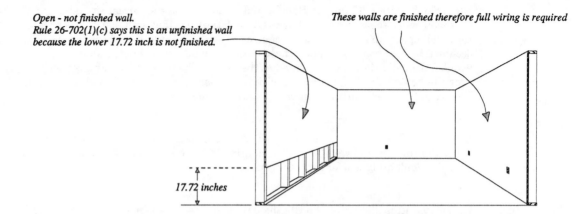

Open - not finished wall.
Rule 26-702(1)(c) says this is an unfinished wall
because the lower 17.72 inch is not finished.

These walls are finished therefore full wiring is required

17.72 inches

(2)   **Each wall or partition** is considered separately. Full wiring is required in all walls and partitions which are finished completely. If only the outside walls of the basement are finished to the floor then only those walls require full wiring as similar rooms on the main floor. Any wall of any basement room where the wall finish stops 45 mm (17.7 in.) above the floor need not be wired except as noted below under "Minimum Basement Plug Outlets Required".

Minimum Basement Plug Outlets Required - Rule 26-702(10)(c) - requires only one plug outlet in an unfinished basement. If there are no walls to divide the basement into two or more rooms or areas the rule is satisfied with just one duplex plug outlet in the whole basement. See also below under "Laundry Plug in Basement".

Note -   **Basement partitions** - unfinished - studs only

If there are no partitions to divide the basement into two or more areas AND the lower portion of the outside basement walls are not finished except as described above AND there is no laundry facility in the basement, this rule is satisfied with just one duplex plug outlet in the whole basement.

Caution - Rule 26-702(10)(c) refers to "areas" not rooms. The rule says "at least one duplex receptacle shall be provided in any unfinished basement area." Unfinished partitions consisting of studs only, can and do divide the total basement floor space into two or more areas and each of these areas is required to have at least one duplex plug outlet. Note too that Rule 26-702(2) also refers to a "room or area" and says that both must be treated equally. There may be differences of opinion on this interpretation, therefore you should check with your local inspector before proceeding.

Laundry Plug in Basement. - If laundry facilities are located in the basement the plug outlet for the washer is in addition to the minimum plug outlets described above and it must be on its own circuit.

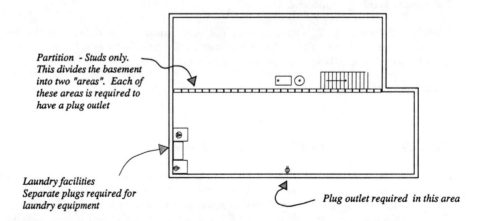

Partition - Studs only.
This divides the basement
into two "areas". Each of
these areas is required to
have a plug outlet

Laundry facilities
Separate plugs required for
laundry equipment

Plug outlet required in this area

Fridge - Duplex receptacle on separate circuit is required except that a recessed clock outlet may also be supplied with this circuit...

Counter Outlets - No point along the back edge of counter may be more than 900 mm (35.5 in) from an outlet.

- 300 mm (11.8 in) or longer counter space requires A plug outlet.

- Adjacent plug outlets may not be on the same circuit.

Dining Area in Kitchen - At least one plug outlet required - must be on separate circuit used for no other load.

See section on services for more detail.

Washroom - Circuit for washer outlet may also supply the receptacle at the wash basin because it is in the laundry room.

Carport or Garage - At least one duplex receptacle is required for each car space. This circuit may also supply the lights and the door opener in the garage or carport.

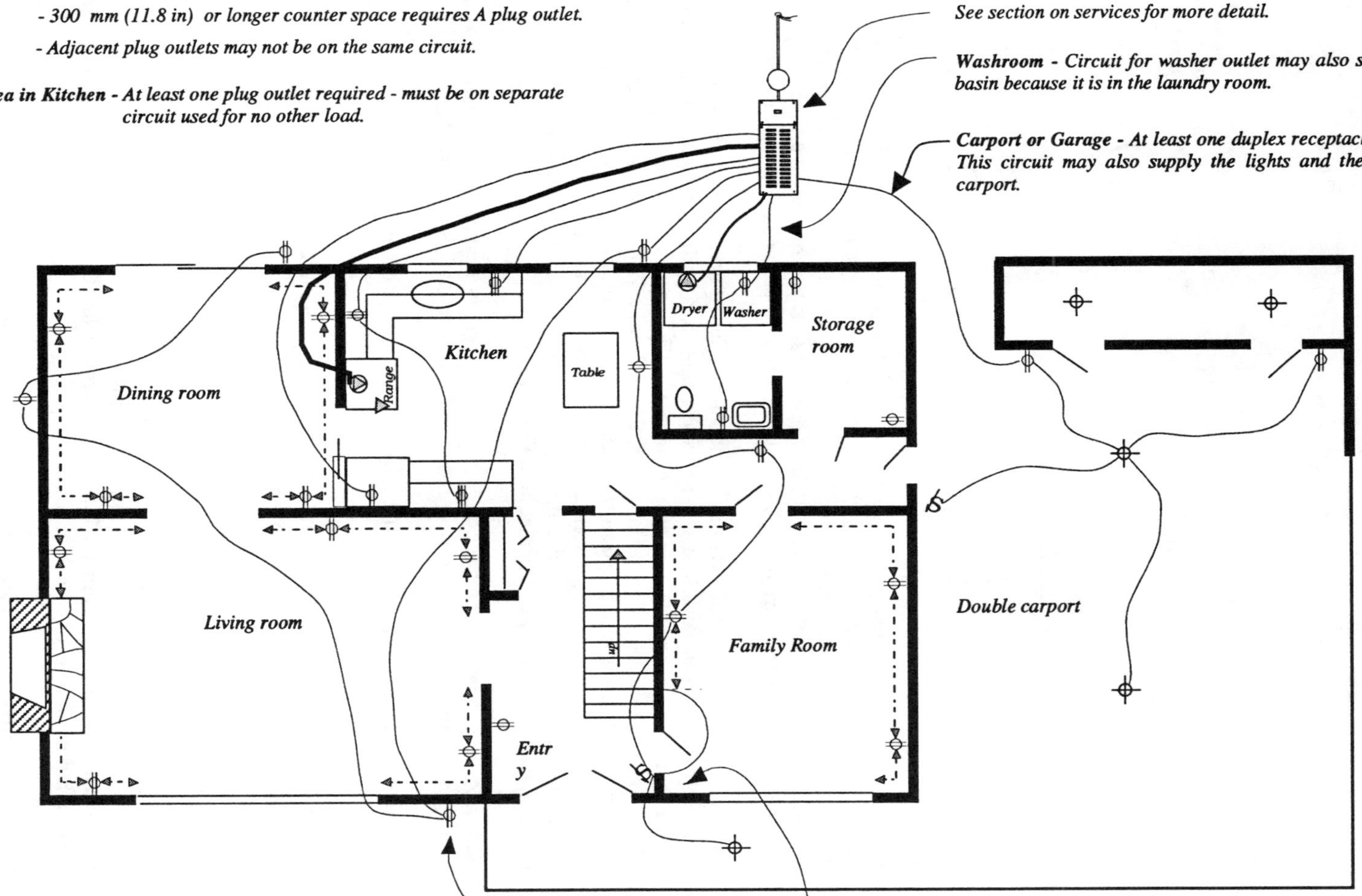

Storage Room & Outdoor Lighting - may not be on the carport or garage plug outlet circuit but lighting in the garage or carport may be supplied with this circuit.

Any wall space 900 mm (35.5 in.) or more in width must have a plug outlet.

Note *This short space requires a plug outlet. The longest wall length is only 600 mm (24 in.) but measured into the corner. as required by Rule 26-702(2) & (5) the length is 1140 mm (45 in.). This is more than 1 m therefore a plug outlet is required - it's the law.*

Outdoor receptacles

- Separate circuit is required.
- Max. 12 outlets per circuit.
- Duplex type required.
- G.F.C.I. type circuit breaker or receptacle must be used.

Double carport

Storage room

Family Room

Washer Dryer

Table

Kitchen

Range

dn

Entry

Dining room

Living room

Notes

- Only plug outlets and a few light outlets are shown.

- Bedrooms are not shown - plug outlets in these rooms are spaced as in a living room or family room.

- Basement wiring is not not shown.

(f)   **Bathroom Plug Outlet** - Rule 26-702(11)(12) requires at least 1 plug outlet in each bathroom.

This outlet must:

A   **Be at least 1 m** (39 in.) away from the bathtub or shower stall. This is a horizontal distance as shown below.

B   **Be adjacent** to the wash basin. The rule requires that this outlet be not above the basin but on either side of it.

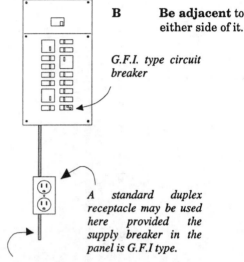

G.F.I. type circuit breaker

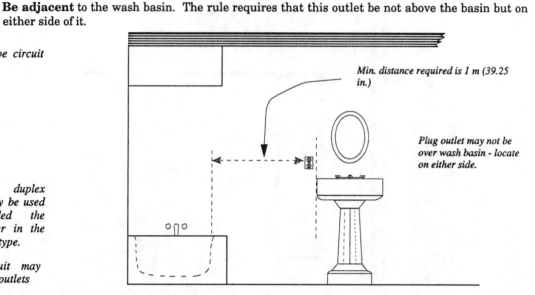

Min. distance required is 1 m (39.25 in.)

Plug outlet may not be over wash basin - locate on either side.

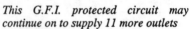

A standard duplex receptacle may be used here provided the supply breaker in the panel is G.F.I type.

This G.F.I. protected circuit may continue on to supply 11 more outlets

C   **Rule 26-700(13)** is a new subrule. It requires GFI protection for plug outlets in **ALL bathrooms and washrooms.** It may be either a G.F.I. type plug receptacle; or it may be an ordinary duplex plug receptacle which is supplied from a special type circuit breaker (G.F.C.I.) in the service panel. This circuit breaker is called a Class A ground fault circuit interrupter. These breakers mount in the service panel as ordinary breakers. If you are using an older type circuit breaker panel make sure they are available for the particular service panel you are using.

This new subrule is very broad in its scope when applied according to the definition for bathroom and washroom. Those definitions do not really refer to a bathroom or to a washroom but to rooms which contain a tub or a shower or a washbasin. The room **contains** this appliance. It could be any room in the house. For example, a large bedroom with a tub is included in this definition. The fact that there is also a bed in the same room is irrelevant. This means that bathroom rules will apply wherever any appliances, such as a tub or shower are being installed. Washroom rules apply wherever there is a washbasin; any room will do. Any plug outlets within 3 m (118 in.) of a bathtub, shower or washbasin must be GFI protected.

D   **May not be a transformer type** plug outlet especially designed for electric razors. These are no longer acceptable. The rules now require GFI protected receptacles.

E   **Circuit required** - This plug outlet may be supplied with any nearby lighting circuit which has 11 or fewer outlets.

F   **Bathroom / Laundry Room Combination** - Rule 26-702(13) - This is an important detail. See under 'Laundry in Bathroom', page 74.

(g)   **Washroom Plug Outlet**

The definition, on page 48 in your code, says a "Washroom means a room containing a wash basin(s) and may contain a water closet(s) but without bathing or showering facilities"

(i)   **Plug outlet Required** - Rule 26-702(11) requires at least one duplex plug outlet adjacent to, ie. next to, but not above, the wash basin.

(ii)   **Circuit Required** - The rules do not require a separate circuit. This outlet may be connected to any nearby lighting circuit.

(iii)   **G.F.I Protection Required** - Yes, this outlet must be GFI protected. Rule 26-700(13) requires GFI protection for all plug outlets within 3 m (118 in.) of a washbasin no matter where the washbasin is located. You may use a GFI circuit breaker in the panel or you may use a GFI type receptacle. In either case the rule permits other plug outlets and lights in other washrooms and bathrooms, bedrooms, hallways etc. to be protected with the same GFI protective device. The maximum number of outlets must not exceed 12. See also above under bathroom plug outlets.

(h)   **Other Rooms or Areas**

(i)   **Storage Rooms** - Rule 26-702(2) - Every finished room, including storage rooms, must have minimum wiring so that no point along the wall is more than 1.8 m (71 in.) from a plug receptacle.

Note   Storage spaces such as areas under the stairways, attics or crawl spaces do not require a plug outlet. In fact it is safer without one, so that appliances must be unplugged before being stored.

(ii)   **Closets, Cupboards, Cabinets etc.**- Rule 26-702(14) prohibits plug outlets in these enclosures except for special cavities built for specific non-heating appliances **designed and certified** for recessed mounting.

(iii)   **Microwave oven** - Rule 26-702(16) specifically permits a microwave oven to be located in a wall cavity. Flush mounted units must be certified for such mounting. Standard, free standing units designed for use on the counter must not be placed in a cavity unless adequate ventilation is provided.

(iv)   **Appliance storage garages** - Rule 26-702(15) & (17) are new subrules which permit a plug outlet in an appliance garage for a **HEATING** appliance. Note the following very specific requirements of this rule for such an appliance garage.

> A   Only one receptacle is permitted; and
>
> B   It must be a single receptacle, not a duplex receptacle; and
>
> C   The garage, complete with one single plug outlet, must be a factory built assembly, not one built on site; and
>
> D   The single receptacle must be controlled by a switch which is operated automatically by the appliance garage door; and
>
> E   The door operated switch must be so arranged that it will remain in the open position until the door is fully open, Rule 26-702(15).

Appliance garage must be factory assembled complete with door operated switch and one single plug outlet.

The Code makers are very concerned with heating appliances being stored away in a cabinet while they are still connected to a power outlet. The concern is an appliance being stored while still in the 'on' position or an appliance which becomes overheated due to a malfunctioning of its control circuit. Such an overheated appliance could very quickly set fire to the combustible construction materials all around it.

This new subrule refers only to **HEATING** appliances. The Code does not define what is included in the term "heating appliance". It would normally mean an appliance, such as a toaster, which is designed to heat something. It is not likely the term would include mixers or grinders which could become overheated through use or malfunction.

This new subrule makes no provision for a plug outlet in an appliance garage for portable **NON-HEATING** appliances normally used in a kitchen. The appliance garage itself is not a problem and the Electrical Code is not concerned if there is one but it is concerned that there not be any plug outlet in that enclosure for a non-heating appliance except as noted above.

Caution - Certification is required - Rule 26-702(17) requires that the appliance garage to be a factory build enclosure complete with door operated switch and single receptacle. Rule 2-022 requires such factory assembled component parts of electrical equipment to have CSA or equivalent certification before it may be sold, connected or used. This garage may not be built on site.

(v) **Dining Room** - Rule 26-702(6)(c) - Dining rooms which are **not part of a kitchen** must be wired as a living room so that no part along the wall is more than 1.8 m (71 in.) from a plug outlet, Rule 26-702(2). For dining or eating areas **forming part of a kitchen**, see under 'Appliance Outlets' below.

(i) **Kitchen Appliance Plug Outlets**

(i) **Types Required** - Rule 26-702(6) & (21)

Duplex Receptacle - all appliance outlets must be duplex type. The single receptacle may be used in only a few special cases.

SPLIT Duplex Receptacles - must be used for all kitchen counter plug receptacles. Single receptacles may not be used in these locations.

This is a standard duplex receptacle, except that it has a small break-away section of metal on the hot (brass colored) side terminal block as shown at left. When this section of metal is broken off it disconnects the two halves of the duplex receptacle from each other so that each half can be connected to a different circuit. They share a common neutral (white) wire.

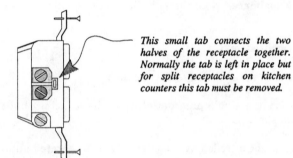

This small tab connects the two halves of the receptacle together. Normally the tab is left in place but for split receptacles on kitchen counters this tab must be removed.

This type of receptacle must be used for all appliance outlets above the kitchen counter work space. Remember, the Code permits two duplex receptacles on one 3-wire circuit. The upper half of each duplex receptacle is on one circuit and the lower half of each receptacle is on the other circuit in the 3-wire supply cable.

(ii) **Adjacent Plug Outlets** - It has been determined, we are not told how, that the average homemaker in an average home is much more likely to use two adjacent receptacles at the same time than two receptacles spaced a bit farther apart. That is why adjacent outlets are not permitted to be supplied with the same 3-wire circuit.

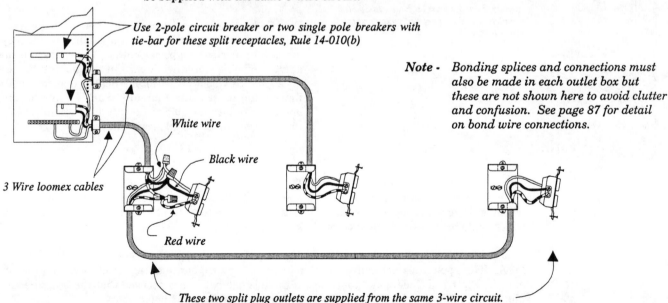

Use 2-pole circuit breaker or two single pole breakers with tie-bar for these split receptacles, Rule 14-010(b)

Note - *Bonding splices and connections must also be made in each outlet box but these are not shown here to avoid clutter and confusion. See page 87 for detail on bond wire connections.*

White wire

Black wire

3 Wire loomex cables

Red wire

These two split plug outlets are supplied from the same 3-wire circuit.

Adjacent counter plug outlets may not be supplied with the same 3-wire cable. The rule says that if only two split receptacles are required they must be supplied with two separate 3-wire cables and circuit breakers. If the counter space is long enough to require 3 split receptacles the third outlet must be located between the first two and would be supplied with a separate 3-wire cable as shown.

Note - Use 2-pole circuit breakers, or 2 single pole breakers with tie-bar, to supply the split receptacles on the counter.

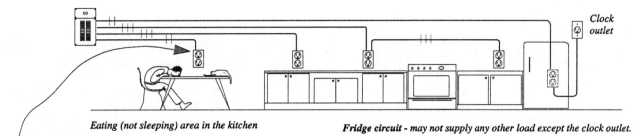

Clock outlet

Eating (not sleeping) area in the kitchen

Fridge circuit - may not supply any other load except the clock outlet.

This plug outlet for the kitchen eating area must be on a separate circuit used for no other load.

Counter plug circuit - may supply only the plug outlets on the counter - it may not supply any other outlets.

Notes (I) **As shown above,** the two split duplex receptacles to the left of the range are on different 3-wire circuits. The outlets on each side of the range are also on different 3-wire circuits. This is to comply with Rule 26-704(3)(c) which says that "adjacent receptacles shall not be connected to the same multi-wire branch circuit".

(2) **Where breakers are used** to supply split receptacles, they must be 2-pole type or they may be two single pole breakers with tie-bar, Rule 14-010, 14-302(b) & Rule 26-708(1).

Where fuses are used Rule 14-010(b) requires that both fuses can be removed simultaneously. A "fuse pull" similar to that required for a range or dryer, is used. Where the fuse panel being used is not equipped with this feature, check with your Inspector for permission to install a separate disconnect switch for these circuits.

(3) **Polarization** - This is an important detail. You will notice the receptacle has a brass terminal screw and the chrome plated terminal screw. Be sure to connect the black, or sometimes the red wire, to the brass terminal screw and the white neutral conductor to the chrome plated terminal screw. The Inspector has a little tester he uses to check this connection without removing any cover etc. If you have connected incorrectly, he will find it.

(ii) **Kitchen** - According to Rule 26-702 & 26-704 kitchen plug outlets are required as follows:

Fridge Outlet - Rule 26-704(2)

A receptacle must be installed for each fridge. It may be a single receptacle. It could be a **split** duplex receptacle but then both circuits of the 3-wire supply circuit would have to end there. One of these circuits would never be used because the receptacle is usually behind the fridge and not readily accessible for other loads. It's best to run a separate 2-wire cable to this location and install a duplex receptacle.

Note - The circuit supplying a fridge outlet may continue on to supply a clock outlet as shown above but it **may not be used** to supply the kitchen hood fan or any other load.

Counter Outlets - Rule 26-702(6)(b)

A sufficient number of receptacles must be installed so that no point along the wall line on the work surface will be more than 900 mm (35.4 in.) from a receptacle. This is measured from the receptacle along the back edge of the counter, as shown below.

Work Surface does not include area of sink, range, fridge or similar appliance.

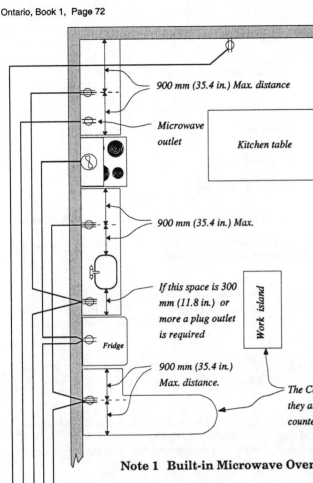

900 mm (35.4 in.) Max. distance

Microwave outlet

Kitchen table

900 mm (35.4 in.) Max.

If this space is 300 mm (11.8 in.) or more a plug outlet is required

Work island

Fridge

900 mm (35.4 in.) Max. distance.

The Code does not require plug outlets on these islands however they are often needed here too. They may be installed below the counter level. They need not be split type.

Eating Area in Kitchen - Rule 26-704(6)(c)

It there is an eating area in the kitchen Rule 26-702(6)(c) requires a duplex plug outlet in that area and Rule 26-704(6) says this plug outlet must be supplied with a circuit used for no other purpose.

Receptacles Required - Each of these work surfaces
must have a plug receptacle. No point along the wall line above the work counter may be more than 900 mm (approx. 36 in.) from a plug receptacle.

Note Any counter space which is 300 mm (approx. 12 in.) long must be provided with one or more outlets. Any counter space less than 300 mm (approx. 12 in.) long is not required to have an outlet.

Receptacle on Range - Rule 26-702(22) - Plug outlets
on the range are not acceptable as alternatives for wall outlets.

Note 1 Built-in Microwave Oven - Rule 26-704(11)

A separate circuit is required for a built-in microwave oven. The plug outlet for a built-in microwave oven must be located in the special cavity built for this oven.

If it is not a built-in microwave oven, i.e., if it stands on the kitchen counter, the rule does not apply. In that case one of the counter outlets may be used to supply this oven.

Note 2 Work Surface - Rule 26-702(6)

As shown above, the counter work space is often divided into several isolated sections. Each of these isolated sections must be considered separately and each must have a split duplex receptacle if it is 300 mm (11.8 in.) or longer. This measurement is along the back wall of the counter space. The reason for this requirement is to make all counter work surfaces properly accessible to appliance receptacles without the supply lines having to cross over sinks, ranges, etc.. Make sure you have a sufficient number of outlets along the counter before covering. It is difficult to add more later after the finish material is in place.

Rule 26-702(8) says that plug outlets may not be mounted facing up in the work surfaces or counters in the kitchen or dining area. The concern is spillage and clean up with a wet cloth. Plug outlets must be mounted on a vertical plane above the level of the work surface so that any spillage in the area will be prevented from entering the outlet.

Note 3 Work Island - Rule 26-702(6)

This is not a problem area. It need not be considered as far as the 900 mm (35.4 in.) distance rule is concerned. An appliance outlet located on the wall at the back of the island, as shown, is considered sufficient for the island.

Note 4 Appliance Storage Garage - Rule 26-702(15) & (17)

Some modern kitchen cabinets provide space inside the cabinets to store the appliances normally left standing on the kitchen counter. There is nothing wrong with this provided the appliances cannot be left plugged in when they are stored in that cavity. This means that there may not be any plug outlets inside that enclosure.

Exception - Rule 26-702(15) & (17) now permit a plug outlet inside of an appliance garage provided power to that outlet is turned off whenever the enclosure door is not fully open. See page 69 for detailed description of this new subrule.

Note 5 **Disabled Persons** - have difficulty using counter plug outlets when they are located along the wall behind the counter. Rule 26-702(7) allows additional split plug outlets to be installed along the front or end wall of the lower cabinet to provide better access for the disabled. It should be noted that the rule does not require these outlets, it simply allows them in addition to the normal outlets. The two subrules that deal with these special plugs are:

A　　　**Rule 26-702(7)** says that these special counter plug outlets are in addition to those shown in the illustration above, **they are not a substitute** for the plug outlets normally required along the counter; and

B　　　**Rule 26-704(5)** permits these additional outlets to be supplied from the 3-wire circuits used to supply the outlets along the wall behind the counter. Actually this rule refers to only one additional split receptacle. The intent, it seems, is to allow one additional split plug on each 3-wire circuit supplying counter plugs.

Split Duplex Receptacles - Rule 26-702(6)(b) all receptacles along counter work surfaces must be split duplex type.

— Single receptacle may not be used.
— Only two such split duplex receptacles may be supplied from a 3-wire circuit except for special outlets, as noted above, under "Disabled Persons".
— This 3-wire circuit may not supply any other load.

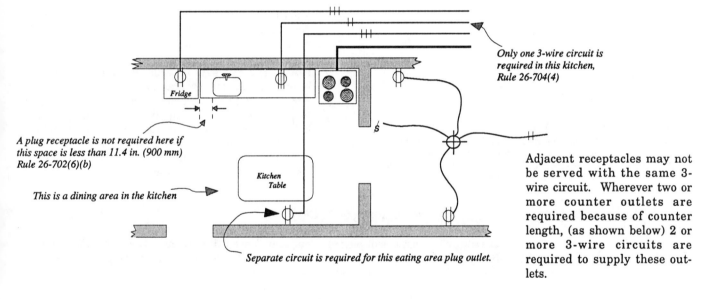

Only one 3-wire circuit is required in this kitchen, Rule 26-704(4)

A plug receptacle is not required here if this space is less than 11.4 in. (900 mm) Rule 26-702(6)(b)

This is a dining area in the kitchen

Separate circuit is required for this eating area plug outlet.

Adjacent receptacles may not be served with the same 3-wire circuit. Wherever two or more counter outlets are required because of counter length, (as shown below) 2 or more 3-wire circuits are required to supply these outlets.

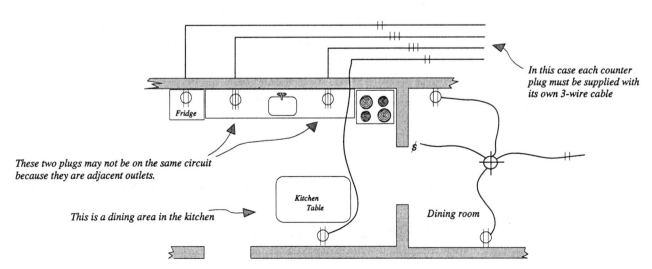

In this case each counter plug must be supplied with its own 3-wire cable

These two plugs may not be on the same circuit because they are adjacent outlets.

This is a dining area in the kitchen

Dining room

(iii)   **Dining Areas**

Kitchen Dining Area - Rule 26-702(6)(c)

If there is an eating area in the kitchen Rule 26-702(6)(c) requires a duplex plug outlet in that area and Rule 26-704(6) says this outlet must be supplied with a separate circuit used for no other purpose. The rule does not require a split receptacle in this location, therefore, the supply cable need be only a 2 wire cable with bond.

Dining Room - Rule 26-702(2)

A dining room which is a separate room and not part of the kitchen must be wired similar to the living room, i.e. no point along the floor line of the walls may be more than 1.8 m (71 in.) from a plug outlet. These outlets may be supplied with any lighting circuit.

(j)   **Laundry Room or Area** - Rules 26-702(10)(a), 26-704(7)

At least one appliance plug outlet must be installed in the laundry room or area.

This laundry room plug outlet:

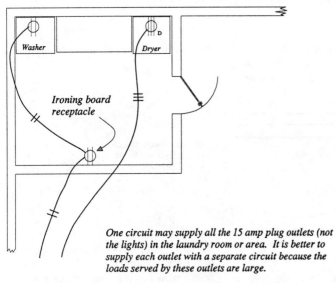

One circuit may supply all the 15 amp plug outlets (not the lights) in the laundry room or area. It is better to supply each outlet with a separate circuit because the loads served by these outlets are large.

-   Must be **duplex type** - a single receptacle is not acceptable in the laundry room. It may be, but does not need to be, a split receptacle.

-   May be at any convenient height in the laundry room.

-   Must be supplied by a circuit used for no other purpose than to supply the one or more duplex appliance plug receptacles in the laundry room or area.

    Note - The rule says we must provide "at least one branch circuit for receptacles" in the laundry room. According to this rule, two or more receptacles in the laundry area could be supplied by one circuit. It is better to install more circuits - one for each outlet. If there is sufficient space to do the ironing in this room you should, the rule does not say must, install a separate circuit for the iron.

(k)   **Utility Room or Area.**- Rules 26-702(10)(b), 26-704(8)

These rules require at least one duplex plug outlet in "each utility room". The term "utility room" is not defined in the Electrical Code nor in the Building Code. It is not referring to the laundry room because that room or area is dealt with separately under another subrule. It is not referring to the porch, or mud room, because that is covered under Rule 26-702(3), see also page 75 for details on porch wiring. It could be the furnace and water heater room but then what on earth would we do with a plug outlet in that room. It would serve no obvious purpose there.

Since there is no definition of the term "utility room" then obviously no one can be sure when he is in that room. Such rooms are difficult to wire properly, but then, it would be just as difficult to prove it was not wired properly.

If the Inspector finds you with a "utility room" - - which has not been properly wired, you could be in serious trouble - - - sort-of.

(l)   **Laundry in Bathroom** - Rule 26-702(13)

Laundry equipment may be located in a bathroom provided the plug outlets for that equipment are properly located.

(1)   **Locate** the washer plug outlet **behind** the washer.

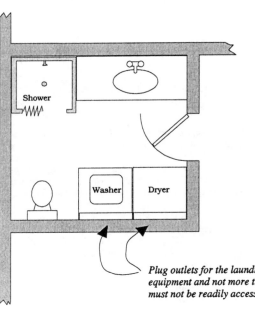

(2) **This washing machine outlet** must not be more than 23.6 in. (600 m) above the floor. The intent here is to make this outlet inaccessible for use with any other electrical appliances which may be used in a bathroom.

(3) **If there is an ironing plug** locate it at least 39.4 inches (1 m) from the tub and shower. This receptacle must be GFI type or be protected with a GFI type circuit breaker in the panel.

Plug outlets for the laundry equipment must be located behind the equipment and not more than 24 in above the floor. The outlets must not be readily accessible for use with any other device.

(vii) **Dryer Receptacles** - see under ``Heavy Appliances'', page 93.

(viii) **Freezer Outlet**

The rules do not demand a separate circuit for the freezer but it is a good idea. We are still allowed to think of this one ourselves. The possible loss of a freezer full of meat because someone tripped the circuit and forgot to reset it, makes this a good investment. Rule 26-704(8) requires a separate circuit for a plug outlet in the utility room. It may be this utility room outlet is intended for a freezer.

(m) **Balcony outlet** - Rule 26-702(3) Yes, the Code now requires a plug outlet on each balcony which is 'enclosed'. The Building Code refers to balconies with 'guards around' and others which are 'enclosed'. Enclosed balconies may well become an added living space with activity requiring electric power. The outlet may be supplied with any lighting circuit - maximum 12 outlets per circuit.

(n) **Porch** - the lowly porch, it has finally been discovered and with its discovery comes the requirement for a plug outlet. A plug outlet is required only if the thing is closed in. The Rule does not require the walls to be finished, just enclosed. This outlet may be supplied with any lighting circuit, - maximum 12 outlets per circuit.

(o) **Outdoor Plug Outlets** - Rules 26-702(18) & (19), 26-704(9)

Basic requirement - Rule 26-702(18) - The rules require at least one plug outlet which is "readily accessible from ground or grade level for the use of appliances which, of necessity, are used outdoors".

These outlets must:

- **Be duplex type** - single outlet is not acceptable.

- **Be readily accessible**, which means it must not be necessary to use chairs or ladders to reach the plug outlet. See definition on page 36 in the Code.

- **Be supplied** with a circuit used solely for this one or more outdoor plug outlets.

- Be G.F.I. type or be supplied with a Class A ground fault circuit interrupter in the panel.

This outlet is for decorative lighting.

This outlet may be any height. If it is more than 2.5 m (98.5 in.) above grade it may be supplied by any lighting circuit inside the building. If it is less than 2.5 m (98.5 in.) above grade the outlet must be GFCI type or be supplied with a GFCI type circuit breaker. This outlet may be supplied by the same circuit used for other outdoor plug outlets.

- If this plug outlet is less than 98.5 in. (2.5 m) above ground or grade level. (not deck level) it must be supplied with a GFCI type circuit breaker or use a GFCI type receptacle.

- Other outdoor plug outlets may also be supplied with the same GFCI protected circuit used to supply this outlet.

- This is the preferred location for the required outdoor plug outlet.

- It must be a GFCI type receptacle or be supplied with a GFCI type circuit breaker.

- Other outdoor plug outlets may also be supplied with this GFCI protected circuit.

NOTES and BEWARES.

(1) **Sundeck Plug Outlets** - Rule 26-702(18) & (19) - If the deck is low and the plug outlets are within 98.5 in. of ground or grade level the plug outlets must be G.F.I. protected with either a circuit breaker in the panel or a G.F.I. type receptacle. These outlets must be supplied with a separate circuit used for no other purpose or with the circuit which supplies the other outdoor outlets.

(2) **Higher Sundecks** - Rule 26-702(19) - Where the outlet on the sundeck is more than 98.5 in. (2.5 m) above ground or grade level, it may still be supplied with the same G.F.I. circuit breaker as noted above, but it is not required to be on this circuit. It could be supplied with any nearby lighting circuit.

(3) **Other Outdoor Plugs** - Rules 26-702(19) & 26-704(9) - These rules could be interpreted to apply to all outdoor plug outlets including those **located in a flower garden** for special lighting and plugs located on the outside surface of other out buildings such as, for example, a separate garage. These plugs are "located outdoors" and could therefore be required to comply with the requirements for separate circuit and GFCI protection. Check with your local Inspector

(4) **G.F.C.I. Protected Plug Circuits Required** - The Rules require **separate** G.F.I. protection for the following plug outlets:

    1   **All bathroom plugs** - within 118 in. (3 m) of a bathtub or shower stall, (except washing machine and dryer plugs in a combined bath and laundry room), Rule 26-700(13); and

    2   **All washroom plugs** - within 118 in. (3 m) of a wash basin, (except washing machine and dryer plugs in a combined washroom and laundry room), Rule 26-700(13),; and

    3   **All carport plugs** - See explanation below, under "Carport only Plug Outlets".

    4   **All outdoor plugs** - within 98.5 in. (2.5 m) of grade, Rule 26-702(19).

Things do get complicated don't they.

(5) **Decorative Lighting Outlets (Christmas Lighting)** - These plug outlets need not be G.F.I. protected if they are 98.5 in. (2.5 m) above ground or grade level. If they are 98.5 in. or less above grade level they must be wired as described for any other outdoor plug outlet, Rule 26-702(19).

(6) **Two Wire Cable** - Use only 2-wire cable for this circuit. The G.F.I. will not work if wired with 3-wire cable.

(7) **Maximum Circuit load** - The rule requires at least one circuit for outdoor plug outlets. Where there are two or more outdoor plug outlets these may all be supplied with one circuit provided the total number of outlets does not exceed I2.

WHAT'S A G.F.C.I. FOR? WHAT'S IT DO? - - - - it saves lives.

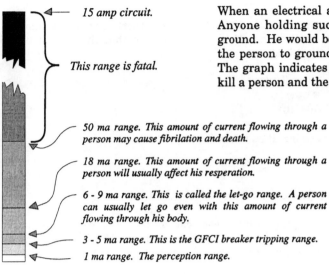

15 amp circuit.

This range is fatal.

50 ma range. This amount of current flowing through a person may cause fibrilation and death.

18 ma range. This amount of current flowing through a person will usually affect his resperation.

6 - 9 ma range. This is called the let-go range. A person can usually let go even with this amount of current flowing through his body.

3 - 5 ma range. This is the GFCI breaker tripping range.

1 ma range. The perception range.

When an electrical appliance is faulty the appliance itself may become energized. Anyone holding such an appliance could become part of that supply circuit to ground. He would become an electrical conductor. The current would flow through the person to ground. Only a very small current is needed to kill a human being. The graph indicates the enormous difference between the small amount needed to kill a person and the large amount available in every 15 amp. circuit in the house.

It would be very expensive to protect all the circuits in the house with these special circuit breakers. What's more, it is not necessary. The code requires only certain outlets to be protected with this special breaker. As indicated above, outdoor plug outlets are among those outlets which must be protected with this special circuit breaker.

The graph also indicates the very low current that a G.F.C.I. will pass to ground. It is designed to trip at a maximum 5 m.a. which is only 0.005 ampere. This means that a person holding a faulty electrical appliance, such as an electric lawn mower or an electric drill, could become an electrical conductor to ground and he could get an electrical shock but the circuit breaker would open the circuit before the current reached a dangerous level.

(p) **Carport only Plug Outlets** - Rules 26-702(20), 26-704(10)

At least one plug outlet must be installed for each car space in the carport.

This outlet must:- **Be duplex type** - single outlet is not acceptable.

Be installed, one in each car space. (The rule does not say 'in' each car space it says 'for' each car space. However, the intent of the rule seems to be that there should be a plug outlet **in** each car space.)

Be supplied with a circuit used solely for these outlets located in a carport except that the carport lighting may also be supplied with this circuit.

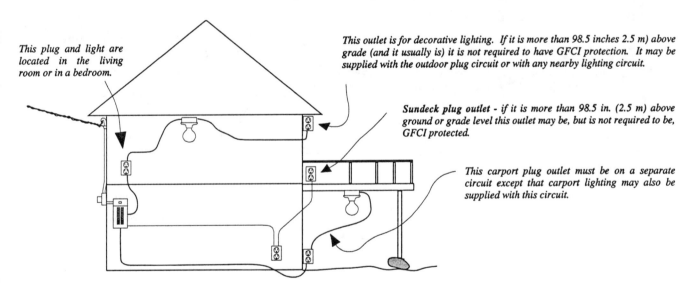

This plug and light are located in the living room or in a bedroom.

This outlet is for decorative lighting. If it is more than 98.5 inches 2.5 m) above grade (and it usually is) it is not required to have GFCI protection. It may be supplied with the outdoor plug circuit or with any nearby lighting circuit.

Sundeck plug outlet - if it is more than 98.5 in. (2.5 m) above ground or grade level this outlet may be, but is not required to be, GFCI protected.

This carport plug outlet must be on a separate circuit except that carport lighting may also be supplied with this circuit.

Outdoor plug outlet accessible from grade must be supplied with a separate circuit used for no other load except other outdoor plug outlets.

Notes **(1) G.F.I. Protection Required?** - Yes, maybe. - Although this is a carport plug outlet it is in fact facing outdoors just as any other outdoor plug does, therefore, Rule 26-702(19) could be applied. This rule says all outdoor plug outlets within 98.5 in. of ground or grade level must be protected with a G.F.I. circuit breaker in the panel or with a G.F.I. type plug receptacle. Check this point with your local Inspector.

(2) Separate Circuit Required - Rule 26-704(10) says carport plug outlets must be supplied with a circuit used for no other purpose except that the carport lights may also be supplied with this circuit.

(q)   Garage Plug Outlets - Rules 26-702(20) & 26-704(10)

At least one appliance plug outlet must be installed in each car space in a garage.

This outlet must:

- **Be duplex type** - single receptacle is not acceptable.

- **Be installed** so that there is a plug outlet in each car space. To be truthful, the rule does not say "in" each car space, it says "for" each car space. However, the intent seems to be that each plug outlet should be located in its own car space.

- **Be supplied** with a circuit used solely for the plug outlets located in the garage except that light outlets and garage door openers may also be connected to this circuit.

Note   G.F.I. protection - The garage plug outlets are not required to be protected with a G.F.l. type circuit breaker.

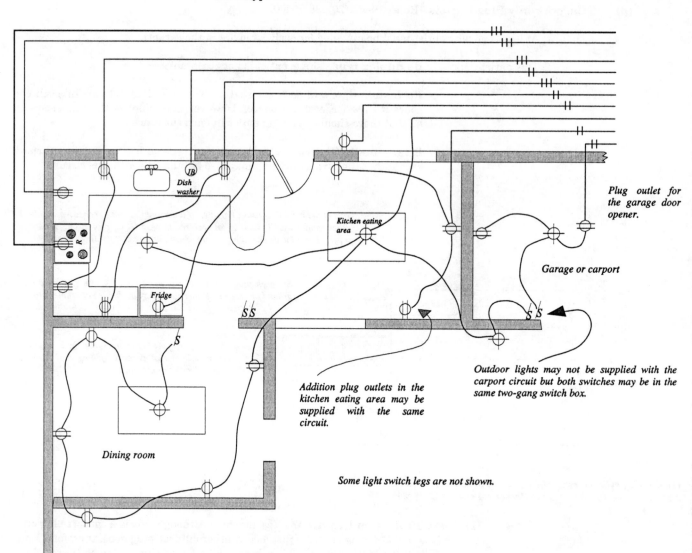

Dish washer

Kitchen eating area

Fridge

Dining room

Plug outlet for the garage door opener.

Garage or carport

Outdoor lights may not be supplied with the carport circuit but both switches may be in the same two-gang switch box.

Addition plug outlets in the kitchen eating area may be supplied with the same circuit.

Some light switch legs are not shown.

(r)   **Junction Boxes**

      (i)   **Accessibility** - Rules 12-3016(1), 12-112(3) - They must remain accessible. This means they may not be concealed in a wall or ceiling or similar place.

      (ii)   **Head Clearance** - Rule 12-3016(2) - Where junction boxes are installed in an attic or crawl space, there shall be at least 900 mm (35.4 in). vertical space to provide access.

      (iii)   **Where to use** - Use junction boxes very sparingly, only where you absolutely have to. Usually all the joints are made in light, switch and plug outlet boxes

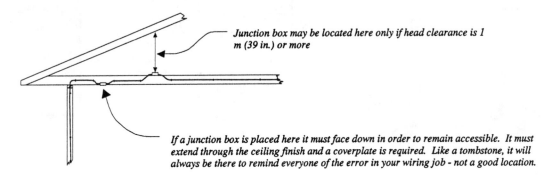

Junction box may be located here only if head clearance is 1 m (39 in.) or more

If a junction box is placed here it must face down in order to remain accessible. It must extend through the ceiling finish and a coverplate is required. Like a tombstone, it will always be there to remind everyone of the error in your wiring job - not a good location.

(s)   **Door Bell Transformer** Rule 16-200, 16-204

      (i)   **Type** - Must be a CSA certified Class II transformer. This is usually die stamped somewhere on the transformer. This Class II label, is very important. It means the transformer is designed so that it will not be a fire hazard even if improperly wired.

      (ii)   **Circuit** - may be connected to any lighting circuit.

      (iii)   **Location** — Watch this one. This transformer must be located somewhere where it will remain accessible. This means it may not be inside a finished wall or ceiling where there is no access.

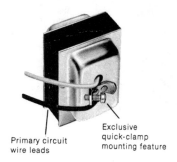

Primary circuit wire leads

Exclusive quick-clamp mounting feature

         Caution - Do not mount this transformer inside the service panel. It may be nippled into the side of the branch circuit panel provided the wall finish is kept back to keep it exposed and accessible.

         The furnace room or basement workshop area is usually a good location for this transformer.

      (iv)   **Cable** - Type LVT cable may be used.

20 TYPE OF BOXES

(a) **Type** - There are many types of boxes available but only a few are in common use today.

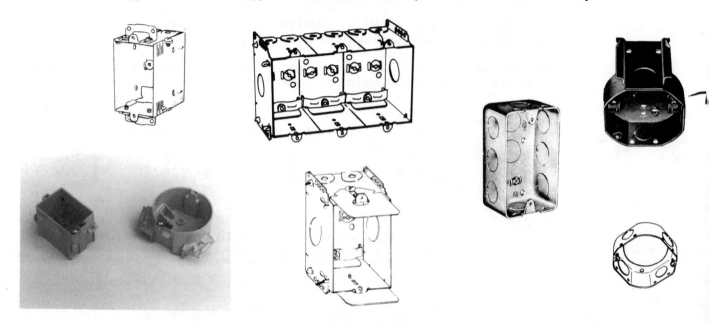

Caution - For bonding metal gang boxes see also under "Bonding of Boxes" on page 83.

(i) **Internal Cable Clamp**

Use the clamp properly. Where the cable enters the back or top knockout, as shown, it must emerge at the lower edge. It is not correct to double it back over the sharp edge, as shown.

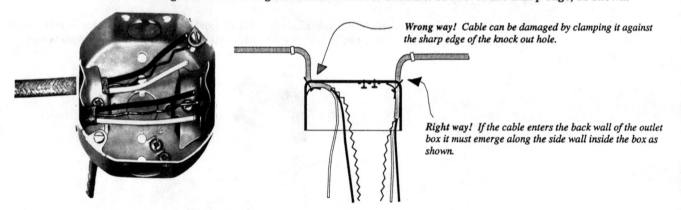

Wrong way! Cable can be damaged by clamping it against the sharp edge of the knock out hole.

Right way! If the cable enters the back wall of the outlet box it must emerge along the side wall inside the box as shown.

(ii) **Saucer Boxes** - The restrictions on the use of this box have been removed. This box may be used anywhere that similar deeper boxes are permitted.

Caution Do not use the center KO hole in this box unless the fixture you plan to connect to this outlet is the simple lampholder type shown.

Most light fixtures use a mounting strap to hold the fixture in place. There is usually a long hollow bolt which runs through the center of the fixture base and into the mounting strap. It's this long hollow bolt that may cause trouble if it extends too far into the shallow box because it is directly in line with the center KO hole in the box. If your supply cable enters the center KO hole it could be seriously damaged with this fixture mounting bolt. For this reason you should never enter a shallow box through the center KO hole. Always use one of the off center KO holes for cable entry. That center KO was not intended for cable entry but for special box mounting and heavy fixture support with a special box supporting bar.

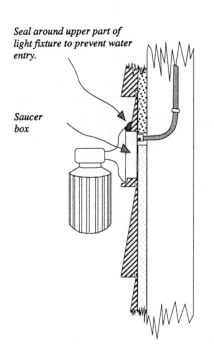

Seal around upper part of light fixture to prevent water entry.

Saucer box

These shallow boxes are often used at front and back doors when the outside wall finish is not smooth. As shown, the box is fastened directly to the outside rough sheathing. Because it is so shallow it need not be recessed into the sheathing. When the finish siding is installed the box may be shifted to match the boards so that it is fully recessed into the outer finish sheathing material. This eliminates the possibility of the outlet box being somewhere on the joint, between two boards, where it is difficult to fit the fixture properly and to seal (weatherproof) the opening around the fixture.

Note - Box Loading - saucer boxes are very shallow, approximately $1/2$ in. deep, and therefore may be used only at the end of a cable run. Only one 2-conductor #14 or #I2 cable may enter this box. This means that you must run to the switch box first then to this light outlet box.

(b)   Box Support - Rules 12-3012, 12-3014

    (i)   Nail-on Type - These are the most practical, they are available in plastic or metal. Note the large ears for fastening. They may be nailed on. If nails are used they should be driven all the way in - not just halfway, then bent over.

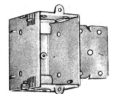

Nails must be located in the corners and not interfere with the conductors or the connectors

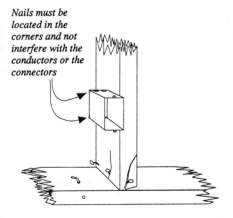

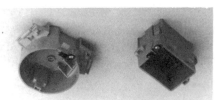

(ii) Sectional Boxes - may be nailed on as shown.

Note -   Rule 12-3012(5) supporting nails may pass through the box only if they are hard against either the ends or back of the box.

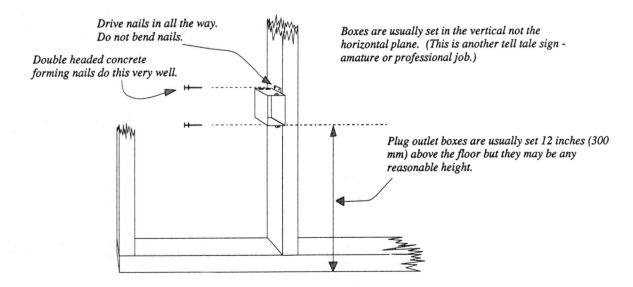

Drive nails in all the way. Do not bend nails.

Double headed concrete forming nails do this very well.

Boxes are usually set in the vertical not the horizontal plane. (This is another tell tale sign - amature or professional job.)

Plug outlet boxes are usually set 12 inches (300 mm) above the floor but they may be any reasonable height.

(iii)   **Metal Gang Boxes** - Rule 12-3012(2)

Metal gang boxes may be supported with a brace, as shown, or with wood backing.

Note -   The revised Rule requires that metal sectional boxes be secured with metal supports or have a header board fixed between the studs. Metal supports are the easiest to use. For metal sectional gang boxes use one of the metal side plates which were removed to form the gang box. Use the side which has the nail-on lugs and fasten one end to the back of the gang box with one of the bonding screws installed from the back of the box. The other end of the brace is nailed to the stud, as shown.

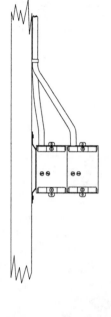

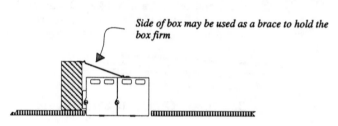

Side of box may be used as a brace to hold the box firm

An alternative method is to fasten the gang box to a wood member installed behind the box. This is usually an unhappy experience because of the many screws protruding through the back of the box.

Caution - For bonding of sectional metal boxes see under "Bonding of Boxes" page 83.

(c)   **Set Flush**

(i)   **Rule 12-3018** - Set all boxes so that they are flush with the finished surface. Pay particular attention to the feature walls of wood paneling. Don't forget to allow for the thickness of the wood strapping.

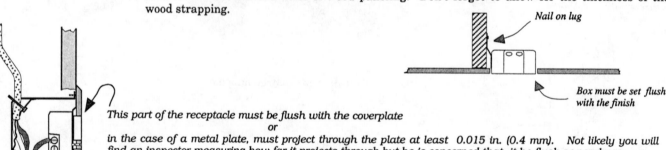

Nail on lug

Box must be set flush with the finish

This part of the receptacle must be flush with the coverplate

or

in the case of a metal plate, must project through the plate at least 0.015 in. (0.4 mm). Not likely you will find an inspector measuring how far it projects through but he is concerned that it be flush or nearly so.

This additional projection required for metal plates is easily obtained if the receptacle is correctly set. A metal plate is much thinner than the bakelite plate so that when the supporting screw is tightened you should have the required projection.

This additional projection for metal plates is important when using older type extension cord attachment caps. These old caps had their terminal screws flush with the front face of the cap. These terminal screws can contact the metal plate and cause a short circuit. Where the metal plate is set back as required such short circuits can be avoided.

(ii)   **Box Extenders** - Approved box extenders must be used where the box is not flush with the finish.

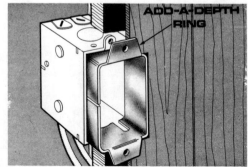

ADD-A-DEPTH RING

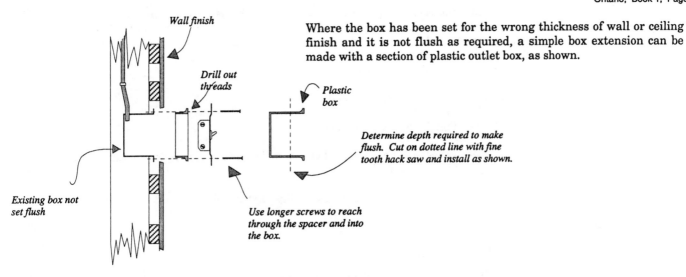

Wall finish

Drill out threads

Plastic box

Where the box has been set for the wrong thickness of wall or ceiling finish and it is not flush as required, a simple box extension can be made with a section of plastic outlet box, as shown.

Determine depth required to make flush. Cut on dotted line with fine tooth hack saw and install as shown.

Existing box not set flush

Use longer screws to reach through the spacer and into the box.

(d) **Bonding of Boxes** - Rules 12-526, 10-400, 10-808(2), 10-906

The bare bonding conductor in the outlet box must be properly connected as shown below.

Note - The bare wire should:

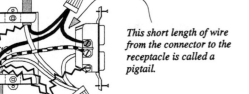

(I) **First connect** to each box as shown.
- In metal gang boxes it connects **to EACH SECTION**. Leave one long enough to loop it around a bonding screw in each section then join them all together.
- In plastic gang boxes it connects to each metal strap inside the box unless these straps are already joined together by the manufacturer.

(2) **Next it connects** to each bare wire entering the outlet box. See under "Conductor Joints & Splices" on page 87 for comments on methods of joining these wires.

(3) **Last, it connects** to the green bonding screw on the base of the plug receptacle.

Pigtails Required - Rule 808(2) & 10-906(6)

This short length of wire from the connector to the receptacle is called a pigtail.

As noted above, where there are two or more ground wires they must be properly joined and connected to both the box and receptacle. They may not be twisted together and wrapped around a bonding screw. Rule 10-808(2) says that only one wire may be fastened to any one bonding screw. If you have two or more bonding wires connect one to the bonding screw in the box but leave it long enough so that the other wires can be spliced to it as shown.

(e) **Wires in Box** - Rules 12-3002(4), 12-3038(1)(c)

(i) **Free Conductor**

At least 150 mm (6 in.) of free conductor must be left in the outlet box to allow joints to be made or fixtures to be connected in a workmanlike manner.

(ii) **Cable Sheath** - Rule 2-108 - Workmanship

The outer cable sheath should not project into the outlet box more than 0.5 in. past the connector. Remove this outer sheath as required before installing the cable in the connector. This sheath is very difficult to remove properly once the cable is actually installed.

(iii) **Box Fill** - Ont Bulletin 12-1-6

Watch that box fill, it's tricky. This otherwise simple problem has been made difficult in the code book. We must count the number of wire connectors with insulated caps which we install in a box. We must subtract the space occupied by these connector caps from the space in the box. Of course, we are still required to note carefully the number of wires entering the box - deducting 1.5 cubic inches for each insulated #14 conductor. The bare conductor does not count. Then, the rules say the switch or receptacle in the box occupies space equal to two conductors. The tables below take all these factors into account.

The following should be carefully noted:

— Pigtails do not count as box fill.

— Boxes may have internal or external cable clamps - it does not matter. The box fill is the same for both.

— Wires from directly connected light fixtures do not count as box fill.

NOTES ON BOX FILL TABLE

(1) Nominal Dimensions

Don't let the nominal box dimensions fool you. These are not the actual box sizes. It should be noted that some plastic (phenolic) box manufacturers keep the dimensions of their boxes close to the nominal. It is better to work from box volume than from its dimensions. Most box manufacturers now mark the cubic volume of their boxes with a die stamp inside the box. Look for these marked boxes.

(2) Various Combinations Given in the Table

In some combinations the Tables allow many more connector caps than could possibly be used for the number of wires in the box. It also shows other combinations where there are not enough connector caps for the number of wires in the box. Choose a combination which will permit you to install at least the number of wires you need and at least the number of connector caps you require. For example, if we are using #14 loomex cable and intend to run a 2-wire supply cable and a 2-wire load cable into a 2 in. deep switch box and plan to make 2 joints in this box the Table says NO! It's too full. We may install only 4 - #14 wires and one connector cap in this box.

Note - only the insulated wires count as box fill - the bare wire does not count. A deeper box is required. The Table shows a 2.5 in. deep box may contain the 5 - #14 wires and the two splices (connector caps) we need. We do not need to provide the extra space as far as the rules are concerned but this is the combination nearest to our needs and it gives us room for a little error in our planning.

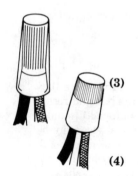

(3) Insulated Cap

The rule says if we use these, we must reduce the number of wires in the box. The Table on page 85 takes all this into account.

Note - this applies to insulated caps of all kinds.

(4) Tape Insulated Joints & Splices

The rule does not mention tape insulated joints or splices - perhaps it is because they occupy less space in the box. If you are using tape to insulate splices. you may add one conductor to the box fill indicated in the table. See also under "Conductor Joints & Splices" on page 87.

If for some reason you have misjudged and find you have more "things" in a box than the rules permit you to have, simply change from insulated caps to soldered and taped splices. These do not count as box fill. See page 87 for soldering instructions.

(5) Bare Bonding Wires

These wires are in the box but are not counted as box fill - Rule 12-3038(1).

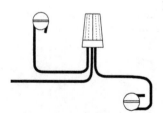

Note - Insulated caps on connectors used to splice the bare bond wires do count as box fill - Rule 12-3038(2) refers to "every" insulator cap. Insulation is really not required on bond wire splices - a crimp-on connector may be used without insulation and these do, not count as box fill.

ACTUAL BOX FILL PERMITTED - RULE 12-3038

Box size and number of sections in the gang — See also Notes on page 84	Actual cu. in. volume of boxes	Using #14 cable this box may contain:	Using #12 cable this box may contain:
Single gang - metal box. Box size 3 x 2 x 1.5 in. deep — See notes page 84	8	One switch or one plug outlets plus: 3 3 2 Wires 0 1 2 Caps	One switch or plug outlet plus: 2 2 Wires 0 1 Caps
2-gang - metal box. Each section is 3 x 2 x 1.5 in. deep	16	Two switches or two plug outlets plus: 6 6 5 5 4 Wires 0 1 2 3 4 Caps	2 switches or 2 plug outlets plus: 5 5 4 Wires 0 1 2 Caps
3-gang - metal box. Each section is 3 x 2 x 1.5 in. deep	24	Three switches or three plug outlets plus: 10 10 9 9 8 Wires 0 1 2 3 4 Caps	3 switches or 3 plug outlets plus: 7 7 6 6 5 Wires 0 1 2 3 4 Caps
Single gang - metal box. Box size is 3 x 2 x 2 or 3 x 2 x 2.25 in deep — See note page 84	10	One switch or one plug outlet plus: 4 4 3 3 2 Wires 0 1 2 3 4 Caps	One switch or one plug outlet plus: 3 3 2 2 Wires 0 1 2 3 Caps
2-gang - metal box. Each section is 3 x 2 x 2 or 3 x 2 x 2.25 in. deep	20	2 switches or 2 plug outlets plus: 9 9 8 8 7 7 Wires 0 1 2 3 4 5 Caps	2 switches or 2 plug outlets plus: 7 7 6 6 5 Wires 0 1 2 3 4 Caps
3-gang - metal box. Each section is 3 x 2 x 2 or 3 x 2 x 2.25 in. deep	30	3 switches or 3 plug outlets plus: 14 14 13 13 12 12 Wires 0 1 2 3 4 5 Caps	3 switches or 3 plug outlets plus: 11 11 10 10 9 9 Wires 0 1 2 3 4 5 Caps
Single gang - metal box. Box size is 3 x 2 x 2.5 in. deep — See notes on page 84	12.5	One switch or one plug outlet plus: 6 6 5 5 4 4 Wires 0 1 2 3 4 5 Caps	One switch or one plug outlet plus: 5 5 4 4 3 Wires 0 1 2 3 4 Caps
2-gang - metal box. Each section is 3 x 2 x 2.5 in. deep	25	2 switches or 2 plug outlets Plus: 12 12 11 11 10 10 Wires 0 1 2 3 4 5 Caps	2 switches or 2 plug outlets plus: 10 10 9 9 8 8 Wires 0 1 2 3 4 5 Caps
3-gang - metal box. Each section is 3 x 2 x 2.5in. deep.	37.5	3 switches or 3 plug outlets plus: 19 19 18 18 17 17 Wires 0 1 2 3 4 5 Caps	3 switches or 3 plug outlets plus: 15 15 14 14 13 13 Wires 0 1 2 3 4 5 Caps
Single gang - metal box. Box size is 3 x 2 x 3 in. deep	15	One switch or plug outlet plus: 8 8 7 7 6 6 Wires 0 1 2 3 4 5 Caps	One switch or one plug outlet plus: 6 6 5 5 4 4 Wires 0 1 2 3 4 5 Caps
2-gang - metal box. Each section is 3 x 3 x 2 in deep	30	2 switches or 2 plug outlets plus: 16 16 15 15 14 14 Wires 0 1 2 3 4 5 Caps	2 switches or 2 plug outlets plus: 13 13 12 12 11 11 Wires 0 1 2 3 4 5 Caps
3-gang - metal box. Each section is 3 x 2 x 3 in deep	45	3 switches or 3 plug outlets plus: 24 24 23 23 22 22 Wires 0 1 2 3 4 5 Caps	3 switches or 3 plug outlets plus: 19 19 18 18 17 17 Wires 0 1 2 3 4 5 Caps
Single gang - plastic box. These boxes usually have their volume clearly marked.	16	One switch or one plug outlet plus: 8 8 7 7 6 6 Wires 0 1 2 3 4 5 Caps	One switch or plug outlet plus: 7 7 6 6 5 5 Wires 0 1 2 3 4 5 Caps
Single gang - plastic box. These boxes usually have their volume clearly marked.	18	One switch or one plug outlet plus: 10 10 9 9 8 8 Wires 0 1 2 3 4 5 Caps	One switch or plug outlet plus: 8 8 7 7 6 6 Wires 0 1 2 3 4 5 Caps

Box size	Cubic inch capacity	OCTAGONAL BOXES								Maximum combination wires and caps using #12 wire							
		Maximum combination wires and caps using #14 wire															
Metal octagonal box 4 x 1½in. deep.	15	**Wires**	10	10	9	9	8	8		8	8	7	7	6	6		
		Caps	0	1	2	3	4	5		0	1	2	3	4			
Metal octagonal box 4 x 2 1/8 in. deep.	21	**Wires**	14	14	13	13	12	12	11	11	12	12	11	11	10	10	9
		Caps	0	1	2	3	4	5	6	7	0	1	2	3	4	5	6
Shallow saucer type box. Limited use rule 12-3002(4)	5	**ROUND METAL BOXES**															
		Wires	3	3	2						2	2					
		Caps	0	1	2						0	1					
Round plastic box 4 x 1½ in deep.	14	**ROUND PLASTIC BOXES**															
		Wires	9	9	8	8	7	7			8	8	7	7	6		
		Caps	0	1	2	3	4	5			0	1	2	3	4		
Round plastic box 4 x 2 1/8 in. deep.	22	**Wires**	14	14	13	13	12	12	11		12	12	11	11	10	10	9
		Caps	0	1	2	3	4	5	6		0	1	2	3	4	5	6
Round plastic box 4 x 2 5/8 in. deep	28	**Wires**	18	18	17	17	16	16	15	15	16	16	15	15	14	14	13	13
		Caps	0	1	2	3	4	5	6	7	0	1	2	3	4	5	6	7

(f)   **Conductor Joints & Splices** - Rules 4-034(4), 10-808(2) & 12-506(1)

    (i)   **The illustration** below shows a number of different kinds of outlets and the different connection methods required by code in each case.

To Panel

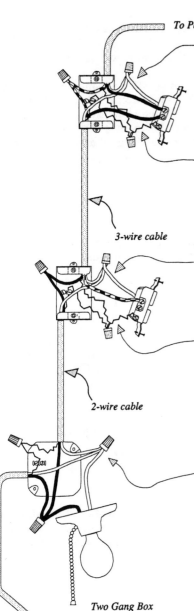

3-wire cable

2-wire cable

Two Gang Box

Neutral wire in 3-wire cable - Rule 4-034(4) requires a pigtail in every case. Without the pigtail both neutral wires would need to be connected to the receptacle. With such a connection anyone removing the receptacle would automatically interrupt the power to other loads downstream.

Hot wires - do not require a pigtail. Both wires may be connected to the receptacle as shown. Connections to split receptacles must be made differently as shown on page 70.

Bare Bond wires - Rules 10-808(2) & 10-906(6) requires a pigtail whenever a circuit continues on to supply other outlets downstream so that the receptacle can be removed without disconnecting downstream outlets from ground. Connect one bonding conductor to the box by taking the shortest route. Do not cut the wire, leave it long enough to make proper splices with other bond wires and a pigtail as shown.

Neutral Wire - This is still 3-wire cable and therefore, Rule 4-034(4) requires a pigtail.

Hot Wires - one connects directly to the receptacle the other is spliced black to black to supply other loads as shown.

Bare Bond Wires - Rules 10-808(2) & 10-906(6) requires a pigtail because the circuit continues on to supply other outlets downstream.

Sometimes an electrician will leave longer lengths of free conductor . He will complete the splice with a connector cap. Later, when he installs the receptacle he will simply loop the bare wire around the bonding screw . This eliminates the need for a pigtail and it is just as acceptable.

Neutral Wires -Pigtails are required here because this is a pull-chain type light fixture which has provision for terminating only one neutral wire, therefore pigtails are required. Ordinary keyless light fixtures usually have provision for terminating two neutrals and two hots. Where such fixtures are used pigtails are not required unless you have more than two neutrals or two hots. Using pigtails to make any of these connections may not always be required by Code but it is still the much better way because it eliminates a lot of strain on the fixture or receptacle termination points.

Hot Wires - See under neutral wire above.

Bond Wires - The fixture used does not require bond wire connection, therefore, the connections may be made as shown .

Neutral Wires - A pigtail is required to join the three neutrals and make the connection to the receptacle.

Hot Wires - A pigtail is required to connect the two devices and to splice the two hot conductors together. Make sure the hot conductors are connected to the brass coloured screws.

Bare Bond Wires - The bare bond wire in the supply cable should be left long enough to connect to the bonding screw in each box then continue on to a wire connector where all the other bare wires are spliced together. A pigtail is required here to connect the receptacle. Do not forget to connect the bond wire to each section of a multi-section metal gang box. One piece metal gang boxes require only one bond connection but multi-gang boxes consisting of a number of sections must have each section connected to the bond wire as shown.

(ii)   **Joints & Splices In Boxes** - Rule 12-506(1)

Joints may be made only in outlet boxes or junction boxes. Junction boxes should be used very sparingly because you can get into more trouble with the Inspector when he finds them. Junction boxes may not be buried in the walls or ceilings.

(iii)   **Solder or Mechanical Joints & Splices** - Rules 10-808(2), 10-906(2), 12-112

Solder - This is probably the best possible method of splicing circuit conductors.

It takes longer to make solder joints. Use a non-corrosive paste, usually 50/50 solder (50% tin 50% lead) for easy flowing and a minimum amount of heat. Then apply scotch electrical tape. Build up a layer of tape equal to the insulation thickness of the conductor. This is needed, not for dialectric strength, but for mechanical protection.

Be sure to melt solder on wire.

Caution -   Do not solder **bonding** conductors - crimp-on type may be used. Taping is not required.

Don't try to cheat if you want to live to a ripe old age. If you are using the soldering method, use it, don't think you can get away with just twisting the wires together, then taping them without soldering the joint - your Inspector will find out.

Crimp-on -   These are good if properly installed. Don't just gimble the connector with your pliers or side cutters and hope the Inspector will not see it. That is poor workmanship and is rejectable according to rule 2-108. Use an approved crimping tool or use the twist-on type wire connectors.

WT-100M also installs STA-KON Terminals and splices #22 to #10 AWG.

Thomas & Betts Ltd.

Twist-on Insulator Caps - There are a number of different types and sizes of twist-on wire connectors available. You must use the correct size to make a good electrical connection. Check the marking on the carton to determine the number and size of wires permitted in each connector.

(g)   **Junction Boxes**

Accessibility - Rules 12-3016(1), 12-112(3) - They must remain accessible. This means they may not be hidden inside a wall or ceiling or similar place.

Head Clearance - Rule 12-3016(2) - Where junction boxes are installed in an attic or crawl space, there must be at least 900 mm (35.4 in.) vertical space above this box to provide access for maintenance.

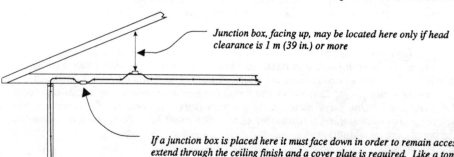

Junction box, facing up, may be located here only if head clearance is 1 m (39 in.) or more

Where To Use - Use junction boxes sparingly, only where you absolutely have to. Usually all your joints are made up in light, switch or plug outlet boxes.

If a junction box is placed here it must face down in order to remain accessible. It must extend through the ceiling finish and a cover plate is required. Like a tombstone, it will always be there to remind everyone of the error in your wiring job - not a good location.

21   LIGHTING FIXTURES

(a)   **Boxes Flush?** - Rule 12-3018

Check first if the boxes worked out flush with the wall or ceiling finish. If not, see page 82 for box extenders.

(b)   **Connections**

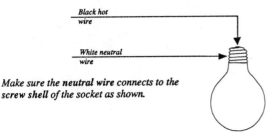

Black hot wire

White neutral wire

Make sure the neutral wire connects to the screw shell of the socket as shown.

Rules 30-314, 30-602 - When connecting light fixtures be sure to connect the neutral white or grey wire to the screw shell of the lamp holder and the black or hot wire to the center pin. The threaded portion of the lamp base must not be energized because of the danger of shock to a person replacing a lamp.

(c)   **Bathroom Light Switch** - Rules 10-400, 30-326, 30-500

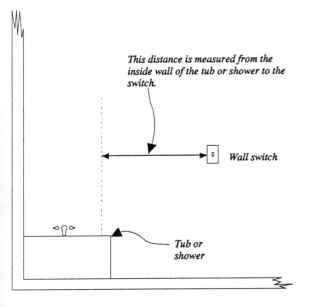

This distance is measured from the inside wall of the tub or shower to the switch.

Wall switch

Tub or shower

Light Fixtures - may be pendant type, such as swag lamps, provided that all the metal on the fixture is properly grounded.

NOTE - The metal chain on chain-hung fixtures may not be used to ground the fixture - Look for fixtures which have a separate grounding conductor run (threaded) through the chain.

All switches must be kept out of reach of a person in a tub or shower, Rule 30-326(3). The Appendix for this rule, on page 604 in the Code, says this means they must be at least 39.4 in. (1 m) away from a tub or shower.

Heat Lamps - Rules 30-200, 62-110, - Like any other recessed light fixture, bathroom heat lamps can be a very real fire hazard if improperly installed.

Care should be taken to locate the fixture away from the door so that it cannot radiate heat directly onto the upper edge of the door when it is left in the open position, see page 55. The rules do not specify any distance, however, a safe distance may be at least 12 inches from the heat lamp to both the door and shower rod. The reason for this is that any clothes or towels left hanging on the door or on a shower rod may be too close to the fixture and could become overheated and cause a fire.

(d)   **Fluorescent Fixtures** - Rule 30-312

Where this type of fixture is mounted end to end in a continuous row as in valance lighting the loomex cable (NMD-90) may enter only the first fixture. It must enter the fixture so that it need not run past the ballast. The connection from there to the other fixtures must be made with type A-18, GTF, R90 or similar types of wire. Care should be taken to see that all the fixtures in the row are properly grounded both for safety and for satisfactory operation.

(e)   **Basement Lighting Fixtures** - Rule 30-326(2)

Pull chain type fixtures must have an insulated link in the chain or have an approved insulating cord when used in the basement or similar area.

(f)   **Low Ceiling-** Rule 30-318

Where fixtures are installed in a crawl space or attic, where there is less than 2.1m (82.7 in.) headroom, the fixture shall be flexible type or be guarded.

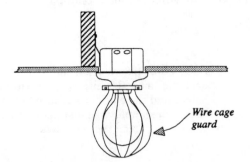

Wire cage guard

(g)   **Closet Fixtures-** Rule 30-204(2)

Fixtures may not be of pendant type. They should be located away from any possible contact with stored items in the closet. See diagram page 55.

22 PLUG RECEPTACLES

(a)   **Boxes Flush'?** - Rule 12-3018

Check first if the boxes worked out flush with the wall or ceiling finish. Check especially the outlets in feature walls. If they are not flush, see page 83 for box extenders.

(b)   **Polarization of Plug Receptacles**

This is an important detail. You will notice the receptacle has a brass terminal screw and a chrome plated terminal screw. Be sure to connect the black or sometimes the red wire to the brass terminal screw and the white neutral conductor connects to the chrome plated terminal screw. The Inspector has a little tester he uses to check this connection without removing any cover, etc. If you have connected incorrectly he will find it.

(c)   **Type** - Rule 26-700(2) and Diagram 1, page 473 in the code

Polarized type receptacles must be used for all plug outlets except clock outlets.

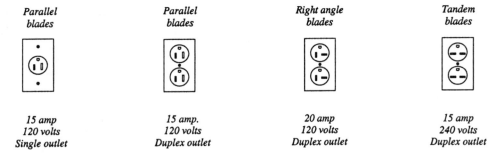

Parallel blades	*Parallel blades*	*Right angle blades*	*Tandem blades*
15 amp 120 volts Single outlet	*15 amp. 120 volts Duplex outlet*	*20 amp 120 volts Duplex outlet*	*15 amp 240 volts Duplex outlet*

20 amp. 120 volt plug receptacles are not interchangeable with 15 amp. as the illustration shows. These should be wired with #12 (20 amp) wire and may be protected with a 20 amp. breaker or fuse.

15 amp. 240 volt plug receptacles are non-interchangeable with the 120 volt receptacles. Single pole circuit breakers used to supply these outlets must be equipped with a tie-bar connecting their operating handels together or use a two-pole breaker. Fuses must have switch or common pull arrangement as used for water heater or dryer, Rule 14-010, 14-302.

Split Receptacle - Note the connection and circuit breakers required for split receptacles. See page 70.

(d)   **Grounding & Bonding** - Rules 10-808(2), 10-906

The bare wire in each outlet box connects first to the box, then to the plug receptacle in every case as shown. In the case of sectional metal boxes the bare wire must connect to the bonding terminal in each section.

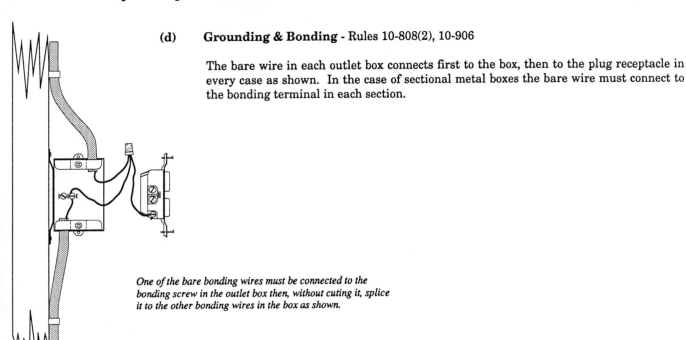

One of the bare bonding wires must be connected to the bonding screw in the outlet box then, without cuting it, splice it to the other bonding wires in the box as shown.

(e)   **Bathroom Plug Receptacle** - Rule 26-702(11)(12) & (13)

Bathroom plug outlets must be either:

(i)   **G.F.I. type** receptacle; or

(ii)   **A standard type duplex plug receptacle** provided it is supplied with a G.F.C.I. type circuit breaker at the panel. This breaker is called a Class A Ground Fault Circuit Interrupter. See also page 68.

(f)   **Outdoor Plug Receptacles** - Rules 26-704(9), 12-3022, 26-706

Receptacles exposed to the weather shall be equipped with spring loaded or threaded cover plates to prevent moisture from entering.

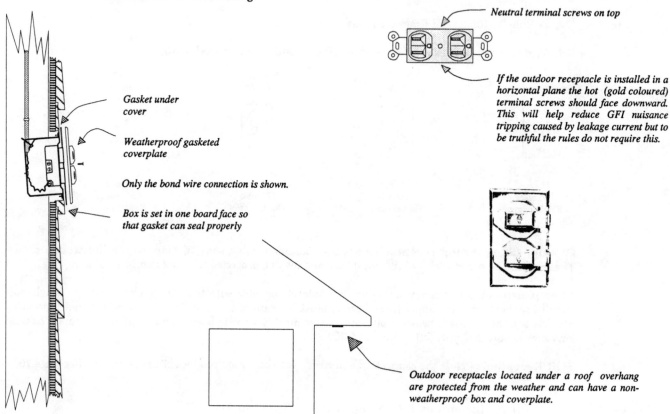

Neutral terminal screws on top

Gasket under cover

Weatherproof gasketed coverplate

Only the bond wire connection is shown.

Box is set in one board face so that gasket can seal properly

If the outdoor receptacle is installed in a horizontal plane the hot (gold coloured) terminal screws should face downward. This will help reduce GFI nuisance tripping caused by leakage current but to be truthful the rules do not require this.

Outdoor receptacles located under a roof overhang are protected from the weather and can have a non-weatherproof box and coverplate.

(g)   **Garden Outlets**

Where outdoor outlet boxes are free standing as in a garden area for decorative lighting the outlet must:

- Be in a weatherproof box - use an F.S. box or equal.
- Be equipped with cover plates held in place with 4 screws.

- Be very well grounded.
- Be above grade.
- These plug outlets must be G.F.I. protected with a special breaker at the panel or you may use a G.F.I. type receptacle. See also under "Outdoor Outlets" on page 75.

The supply cable may be NMWU #14 - depending on the length of run. Mechanical protection must be provided where it runs out of the trench and into the F.S. box.

Rigid metal conduit may be used to protect the cable and to support the box by using a horizontal bend in the trench to hold the conduit upright.

23 HEAVY APPLIANCES - RANGES, DRYERS, GARURATORS ETC.

(a) **General Information**

(i) **System Capacity** - Rules 12-3038 & 8-108 - Before installing additional electrical loads, such as a range, furnace, etc. to a new or an existing service, make sure there is sufficient space in the panel for the branch circuit breakers required to serve the new load. Connecting two or more branch circuits to one fuse or breaker is not approved. Often it is necessary to replace only the branch circuit panel, not the service itself. In any case, the current carrying capacity of the service conductors must, in no case, be less than 60 amperes, Rule 8-200(1)(b)(ii). See also Table of Service Sizes, page 9 of this book.

(ii) **Control** - Rule 14-010(b) - requires that all 240 volt appliances, such as a range etc., must be provided with a device which will simultaneously disconnect both of the hot conductors at the point of supply.

Circuit Breakers require a tie-bar to fulfill this requirement. Rule 14-302(b)(i).

(iii) **Length of Run** - Any length up to 30 m (98 ft.) is acceptable. Longer runs may be acceptable too but they suggest there is a problem with the service location.

(iv) **Mechanical protection** - Rule 12-518 - requires mechanical protection for loomex cable where it is run on the surface of a wall etc. and within 1.5 m (59 in.) from the floor. To comply with this rule, most Inspectors require a flexible conduit installed over the loomex cable supplying a furnace, garburator, etc.

(v) **Cable Strapping** - Rule 12-510 - Cable must be properly strapped within 300 mm (12 in.) of cable termination and every 1.5 m (59 in.).

(vi) **Staples and Straps** - used to support cables must be approved for the particular cable involved.

Note Where cables run through holes in studs or joists they are considered properly strapped. See also page 45 for details on strapping requirements.

(vii) **Grounding** - Rules 10-400, 10-404 - All equipment must be adequately grounded. The bare wire in the supply cable must connect to the bonding screw in the branch circuit panel and to each appliance, using the bonding screw or the bolt provided. It is not enough to wrap the wire around a cable connector or cover screw. Too much depends on a good connection

(b) **Garburator** - Rules 28-106(1), 28-200, 28-600

You will require a separate two wire cable to supply the garburator. A #14 copper cable is adequate in most cases.

The supply cable should run into the control switch outlet box above the counter then down to the motor. You will require flexible conduit to protect the cable from a point inside the wall to the connector on the garburator.

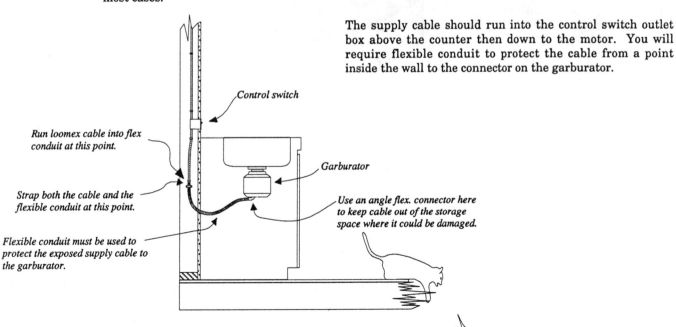

Control switch

Run loomex cable into flex conduit at this point.

Strap both the cable and the flexible conduit at this point.

Flexible conduit must be used to protect the exposed supply cable to the garburator.

Garburator

Use an angle flex. connector here to keep cable out of the storage space where it could be damaged.

(c)   **Dishwasher** - Rules 28-106(1), 28-200

This is a motor and heating load. These usually operate at 120 volts, therefore, a 2-wire cable is required. Unless you are absolutely sure your dishwasher can be served with a #14 cable you should install a #12 copper and 20 amp breaker. You will require a separate circuit. This cable may be run directly into the connection box on the dishwasher. Make sure that the bare wire is connected to the bonding terminal.

(d)   **Domestic Ranges** - Rule 8-300 & Ontario Bulletin 26-12-2

Full Range Wiring Required - Rule 26-746(5)(12)(13) says that wiring for a **free standing** range must be installed in every single family house except where a built-in gas range or built-in electric range is being installed.

(i)   **Free Standing Type** - Rules 26-746, 26-748 & Ontario Bulletin 26-12-2

Freestanding electric ranges must be cord connected. The plug receptacle required for this connection is a 3 pole, 4 wire grounding type as shown below.

Note   The rules, and the current bulletins, do not say that when an existing hard wired range is being replaced that it must then be supplied with a flexible cord as described above. If you have such an installation you should check first with your local Inspector.

FREE STANDING RANGE

Cable size	#8NMD90 copper
Outlet box size	4 $^{11}/_{16}$ x 4 $^{11}/_{16}$ x 2 $^1/_8$
Plug receptacle rating	50 amp.
Rating of fuses or breakers	40 amp. each

Note - Rule 12-3012(3) says this outlet box must be fastend to a solid member directly behind the box or be supported on two of its sides. Where possible locate the box in the stud space behind the range so that it can be **secured to both the plate and a stud**

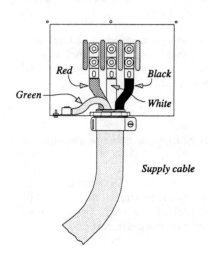

Supply cable

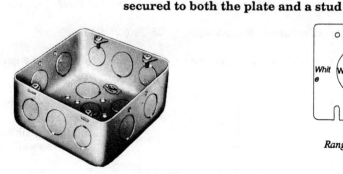

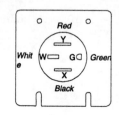

Range receptacle

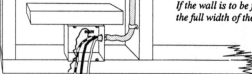

Secure a short length of 2 x 4 to the range or dryer box then secure the combination to the bottom plate as shown.. Once the gyprock is in place on both sides of the studs the box will be held firmly in place.

If the wall is to be finished on one side only the horizontal section of 2 x 4 above the box must then extend to the full width of the stud space and be firmly fixed to the stud on each side of the box.

Notes   There are 3 rules to watch for:

(1)   **The range** outlet box must be located very near the mid-point on the wall behind the range,

AND,

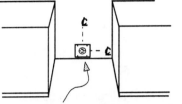

(2) **This range outlet** box must not be higher than 130 mm (5.1 in.) above the floor to the center of the outlet box,

AND,

(3) **This range outlet box** must be carefully positioned so that when the receptacle is finally installed, the ground pin will be either on the right hand or the left hand but not at the top or the bottom.

Set box 4 inches (100 mm) floor to centre. Ground pin must be in the 3 o'clock or the 9 o'clock position but may not at the 12 o'clock or 6 o'clock positions. Make sure the front cover mounting screws, on the box, are in the correct position to allow the receptacle to be mounted properly.

(ii)   Drop-in Type

This is a conventional range but it is not free standing. It is fitted into the kitchen cabinets.

Cable size .... #8 NMD90 copper
Outlet box size .... box not needed
Flexible conduit size .... ³/₄ inch
Rating of fuses or breakers .... 40 amp.

This unit is not cord connected. The #8 NMD90 cable may be run directly into the connection box.

Note - The supply cable must be protected with 3/4 inch flexible conduit for the last 3 feet or so at the range and where it may be subject to mechanical damage.

(iii)   Built-in Type - Rule 26-744

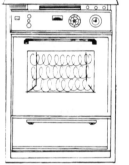

Main cable size .... #8NMD90 copper
Flexible conduit size .... ³/₄ inch
Cable to oven size .... #10 NMD90
Flexible Conduit size .... ¹/₂ inch
Rating of fuses or breakers .... 40 amp. each

Note - The above sizes are sufficient for the average 12 kw. range

Branch Circuit protection - Most range units are equipped with overcurrent protection (fuses or mini-breakers) for the individual elements. If it is not provided as part of each unit, it must be installed by the customer. Rule 26-744. This usually consists of a 6 circuit fuse panel. The branch circuit wires to the elements are normally provided by the manufacturer but they must be connected to the fuse panel by the installer. Each unit also has a length of supply cable attached.

#10 copper cable. This can be loomex type cable. The junction box must be located so that this cable need not be longer than 25 ft. Flexible conduit is usually required to protect this part of the cable to the junction box.

The dotted line is an alternative wiring method using two junction boxes.

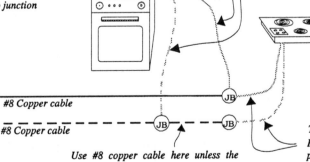

To 40 amp fuses or breakers in the panel.

#8 Copper cable

#8 Copper cable

Use #8 copper cable here unless the cooking top is rated 7200 watts or less.

These cables are part of the cooking top. Flexible conduit is usually required to protect this part of the cable to the junction box.

Wiring to Units - Rule 26-744(2)&(3) - The electrical load in the cooking tops is usually greater than in the oven. Therefore, the junction box is located near the cooking tops so that the supply cable provided by the manufacturer is long enough to run directly into it as shown. The #10 cable to the oven can also be run into this junction box.

Note   Rule 26-744(2)&(3) - The tap cable to the oven must not be longer than 7.5 m (24.6 ft.). Where the units are too far apart for a 24.6 ft. cable, a second junction box may be installed (shown by dotted line) but the #8 supply cable must continue on to the junction box at the cooking tops unless this load is 7200 watts or less.

Cable protection - Flexible conduit is required on these cables for mechanical protection, even though they are in a cabinet below the range. See also Rule 12-518.

Grounding   — If you do not connect the bare ground conductor properly at the panel and at the range, one day the chief cook may not be alive to greet you with a kiss at the end of a busy day.

(e)   **Dryers** - Rules 26-746(3)&(4)

Electrical dryers must be cord connected, they may not be hard wired. Install a 3 pole, 4 wire grounding type plug receptacle behind the dryer location.

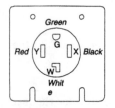

If for any reason a hard wired dryer is being replaced with a new or another used one, the rules require a plug outlet be installed at the dryer.

Cable size ....#10 NMD90 copper
Outlet box size ....4 11/16 x 4 11/16 x 2 1/8 in.
Plug receptacle rating................................30 amp.
Rating of fuses or breakers ....30 amp. each

30 amp Dryer receptacle

Note   Rule 12-3012(3) says the large outlet box required for the dryer must be fastend to a solid structural member directly behind the box or on two of its sides.

These sizes are sufficient for most dryers, i.e. dryers with ratings not greater than 7200 watts.

The bare grounding conductor must be properly connected at the panel and at the dryer to ensure safety to the operator.

Water Heater

Cable size ....#12 NMD90 copper
Flexible conduit size..7/16 inch
Rating of fuses or breakers ....20 amp.
Note - Fuse must be Type P or D...................Rule 14-610

Note - These sizes refer to water heaters with ratings from 3 kw. (3000 watts) up to 3.8 kw. (3800 watts).

(i)   **Circuit** - Rule 26-752(4) - requires a separate circuit with sufficient capacity to carry the maximum possible load that may be connected at one time by the thermostat. Most tanks are provided with 2 - 3 kw. elements, only one of which can operate at one time.   According to Rule 8-302(2) a water heater **may be considered** a continuous load. In that case the supply circuit must not be loaded to more than 80% of the rating of the breaker or fuse. A 3 kw. water heater draws 12.5 amps at 240 volts but the largest continuous load permitted on a 15 amp circuit is 80% or 12 amp. We therefore require 12.5 x 10/8 = 15.6 amp. This means we need a 20 amp breaker or fuse and #12 copper cable.

Note cable strapping. If this is not possible tape the flex cable to the loomex cable at this point.

Use 7/16 inch flexible conduit over the supply cable.

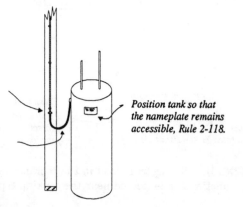

Position tank so that the nameplate remains accessible, Rule 2-118.

Note - If the water heater **is not regarded** as a continuous load or if the water heater load may be treated as permitted for electric heating in Rule 62-114(7) & (8) then a **20 amp circuit breaker and #14 copper cable** may supply a 3000 watt water heater. Check this point with your Electrical Inspector.

Rule 4-034(1) permits a 2-conductor loomex cable to be used to supply a 240 volt load. When this is done the white wire is not used as a neutral but is hot. Wherever this white wire is visible, as in a junction box etc. it must be painted or taped to show a colour other than white, grey or green.

(ii) **Cable Protection** - Rules 12-518, 12-1004 - Protect the hot water tank supply cable with $^7/_{16}$ inch flexible conduit where exposed to mechanical injury.

(g) **Furnace (gas or oil)** - Rule 26-806(1) has been interpreted by some to mean that the supply cable to the furnace must not be with a (two circuit) 3-wire cable, but that it must be a separate two wire cable all the way back to the panel. Check with your local Inspector.

Gas Furnace

Cable size ....#14 copper
Flexible conduit size ....$^7/_{16}$ inch
Fuse or breaker size...............................15 amp.

Oil Furnace

Cable size ....#12 copper
Flexible Conduit size ....$^7/_{16}$ inch
Fuse or breaker size...............................20 amp.

(i) **Disconnect Switch** - Rules 26-806(4)(5)(6), 28-600

A disconnect switch is required for each furnace.

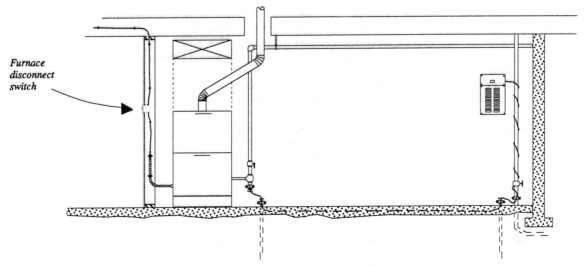

Location - A circuit breaker in a branch circuit panel may serve as disconnect switch provided it is located between the furnace and the escape route.

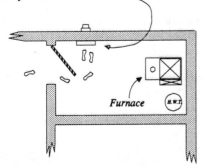

In this case the circuit breaker in the panel is accessible without having to pass the furnace. This is an acceptable location for the disconnect switch

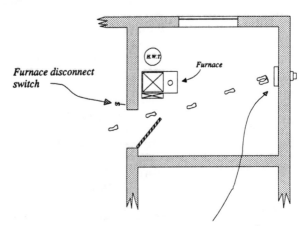

The branch circuit breaker in this panel may not serve as the furnace disconnect switch because the escape route is past the furnace. The furnace disconnect switch must be located as shown.

(ii) **Protection** - Rule 26-802 - The loomex cable is run in a $^7/_{16}$ inch flexible conduit for the last few feet near the furnace where it is less than 1.5 m (59 in.) from the floor or where subject to damage.

(iii) **Grounding** - Rule 10-400 - Be sure to connect securely the bare grounding conductor in the supply cable to the grounding terminal in the furnace junction box.

(iv) **Gas Piping**- Rule 10-406(4) - Natural gas piping supplying the furnace must be bonded to the service grounding electrode. In the illustration on page 97 the service grounding conductor continuous on from the last ground rod to the gas line at the furnace. Do not cut the grounding conductor, simply run it through the ground clamps and on to the last clamp on the gas line.

Caution - Not all gas utilities agree with this bonding business. Some may not want you to bond the gas piping as shown while others require it to be so bonded. This has to do with isolation of the external and internal piping systems and their own system of cathodic protection. In that case leave the bonding off but check first with your gas utility.

(h) **Kitchen Fan**

Cable size ....#14 NMD90 copper
Flexible conduit size ....$^7/_{16}$ inch
Fuse or breaker rating...........................15 amp.

(i) **The kitchen fan** need not be on a separate circuit. It may be on a general lighting or plug circuit but it may **not be** on any special kitchen appliance circuits, Rule 26-704. It is counted as one outlet when determining circuit loading.

(iii) **Cable Protection** - Rule 12-518 - Protect loomex cable with $^7/_{16}$ inch flexible conduit where it is exposed. This includes that portion of the cable where it is exposed inside the cabinet where it may be subject to damage. Flexible cable must terminate in a flex connector at the fan and be strapped or taped to the loomex cable before it emerges from the wall.

(iv) **Grounding** - Rule 10-400 - Because it is near grounded objects it is very important to connect the bare ground wire securely to the ground terminal in the fan connection box.

(i) **Vacuum System** - Rule 26-702(24) requires that if piping for a central vacuum system is installed a plug outlet must be provided for the unit. Rule 26-704(12) requires that this plug outlet must be supplied with a separate circuit used for no other load.

(j) **Hydromassage Bathtub**

(i) **CSA Certified Units** - Make sure the unit you install has a CSA certification label. This may seem like a picky point but in fact, it is extremely important. It becomes convincingly so when you feel that tingly feeling in the rear while soaking in the tub after a hard day.

(ii) **Separate Circuit Required** - Rule 28-106(1) - This is a motor load, therefore, the rules in Section 28 must be applied. For the rough wiring stage run a separate 2-wire loomex cable to the motor location. Later when the unit is actually installed you will need to install a disconnect switch at the motor location.

(iii) **Disconnect Switch Required** - Rules 28-600, 28-602(3)(e).- These rules require a disconnect switch for this motor. This may be a standard wall switch, such as is used to control light outlets. Don't forget to note the ampere rating of this switch - it must be at least 125% as great as the ampere rating of the pump motor.

Note This is not a hot tub. It is a hydromassage unit only. A hot tub usually has an electric heating element as well as a pump.

One more thing - do not forget to provide access to this motor and its disconnect switch for maintenance purposes.

(iv)   **G.F.C.I. Required** - Rule 68-300 - This motor circuit must be protected with a G.F.C.I. type circuit breaker. There is no exception to this rule. Because this is a motor load this special circuit breaker **may not** also supply any other load.

(v)   **Lights and Plugs** - Rule 68-304 - The usual rules for lights and plugs in bathrooms apply to this special bathroom.

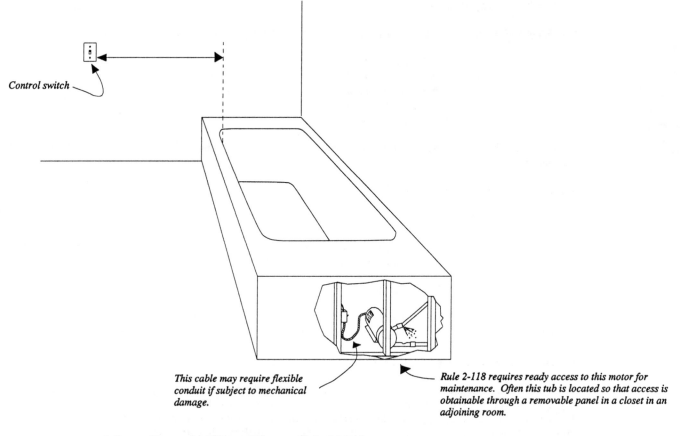

Control switch

This cable may require flexible conduit if subject to mechanical damage.

Rule 2-118 requires ready access to this motor for maintenance. Often this tub is located so that access is obtainable through a removable panel in a closet in an adjoining room.

(vi)   **Control of Pump Motor** - Rule 68-302

         Timer is not Required - Rule 68-302(1) was revised to delete this timer. The rule now requires only a simple wall switch properly rated, (125% of motor full load amps) to control this pump motor.

         Location - Rule 68-302(2) - The control switch must be at least 1 m. (39.4 in.) away from the tub. This is a horizontal measurement from the inside wall of the tub to the control switch. There is an exception - where the control switch is an integral part of a properly certified factory built hydromassage bathtub it may be closer than 1 m from the tub. In fact, in that case, the control switch may be on the tub itself, easily accessible to anyone in the tub. (Do these factory people know somthing the rest of us do not know or do they just use better switches? Maybe they use better bonding methods or maybe they just use better arguments).

24   ELECTRIC HEATING - baseboards.

(a)   **Heat loss calculations** - Ontario Bulletin 62-4-4 - Building heat loss must be calculated according to methods prescribed by Hydro. Duplicate copies of this calculaton must be submitted to Hydro at the time application for inspection is filed.

(b)   **Rough Floor Map Required**

The Electrical Inspector will want a rough sketch of the floor plan of your house showing:

- The location of heaters
- The rating of each heater
- What circuit it is on
- Size of conductors and,
- Rating of breakers

Be sure your sketch is accurate and clear - easy to follow. Your Inspector will want it for that first rough inspection.

(c)   **Branch Circuits** - Rule 62-108

Branch circuits which supply electric heaters may not be used to supply any other load.

2-Wire Cables - Electric heaters are usually connected for 240 volts - no connection to the neutral. Rule 4-034(l) permits a 2-wire loomex cable with one black and one white wire to be used for these loads.

(d)   **Breakers** - Tie-bars - Rule 14-302(b)(i)

Breakers require tie-bars when used to supply 240 volt appliances such as heaters. The tie-bar is used to mechanically connect the operating handles of the breakers so that they operate as one.

(e)   **Circuit Loading** - Rule 62-114(7) & (8)

The rules were changed in the 1990 Code to allow 20 ampere breakers feeding a #14 copper cable to supply 3600 watts of fixed heating load. This is considerably more than was allowed under the old rules. Under the old code the supply cables were never allowed to carry their rated load. This change in the rules was based on the fact that fixed heating loads are just that , they are fixed, they do not change. This is in contrast to other circuits which supply plug outlets or lighting outlets where the total load is constantly changing and is unpredictable.

The following two illustrations show circuit loadings permissible under the rules. The actual wattage rating and the number of heaters used need not be as shown provided the sum of the ratings of all the fixed heaters on the circuit does not exceed the maximum permitted for that circuit.

20 amp Breaker, #14 Copper Cable, Max Load Permitted is 3600 watts

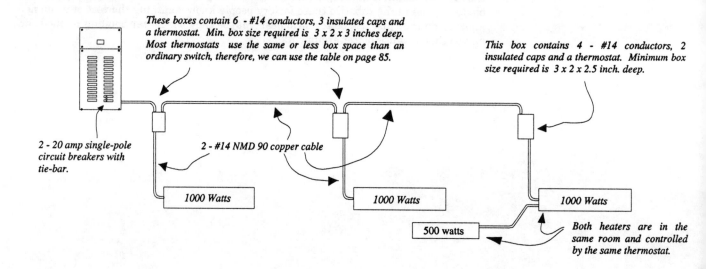

These boxes contain 6 - #14 conductors, 3 insulated caps and a thermostat. Min. box size required is 3 x 2 x 3 inches deep. Most thermostats use the same or less box space than an ordinary switch, therefore, we can use the table on page 85.

This box contains 4 - #14 conductors, 2 insulated caps and a thermostat. Minimum box size required is 3 x 2 x 2.5 inch. deep.

2 - 20 amp single-pole circuit breakers with tie-bar.

2 - #14 NMD 90 copper cable

1000 Watts

1000 Watts

1000 Watts

500 watts

Both heaters are in the same room and controlled by the same thermostat.

Maximum Length - Total length of this circuit run should not exceed 100 feet (approx. 30 m). This is only approximate; only the first part of this cable may be fully loaded. The load is smaller after each heater.

The maximum load permitted with #14 copper cable must not exceed 15 amp. The 20 amp breaker shown protecting this cable is permitted because this is a fixed load. Maximum circuit load must not exceed conductor ampacity multiplied by the circuit voltage, ie. 15 amps x 240 volts = 3600 watts.

Caution - Do not bundle these cables, maintain separation as described on page 43.

Note Under the old rules we would not have been permitted to use a 20 amp breaker; 15 amp was the max rated breaker allowed for a #14 copper cable. With that old arrangement the maximum load permitted was only 2880 watts.

30 amp Breaker, #12 Copper Cable, Maximum Load Permitted is 4800 watts

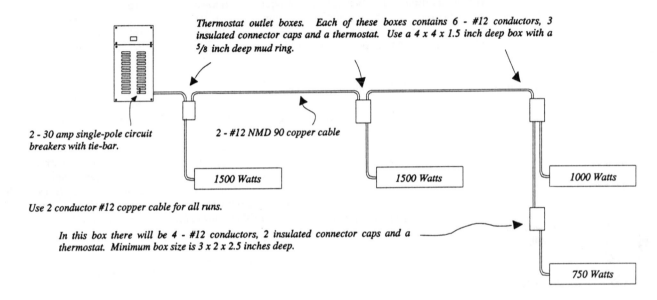

Thermostat outlet boxes. Each of these boxes contains 6 - #12 conductors, 3 insulated connector caps and a thermostat. Use a 4 x 4 x 1.5 inch deep box with a ⅝ inch deep mud ring.

2 - 30 amp single-pole circuit breakers with tie-bar.

2 - #12 NMD 90 copper cable

1500 Watts

1500 Watts

1000 Watts

Use 2 conductor #12 copper cable for all runs.

In this box there will be 4 - #12 conductors, 2 insulated connector caps and a thermostat. Minimum box size is 3 x 2 x 2.5 inches deep.

750 Watts

Maximum Length - Total length of this circuit run should not exceed 100 feet (approx. 30 m). This is only an approximate maximum length because the full load will appear only on the cable to the first heater. After each heater the load becomes less.

The maximum load permitted with #12 copper cable must not exceed 20 amp. The 30 amp breaker shown protecting this cable is permitted because this is a fixed load. Maximum circuit load must not exceed conductor ampacity multiplied by the circuit voltage, ie. 20 amps x 240 volts = 4800 watts.

Note Under the **old rules** we would not have been permitted to use a 30 amp breaker to supply a #12 copper conductor for a fixed heating load. Maximum breaker rating, then, was 20 amp for a #12 copper cable supplying a fixed heating load. With that arrangement, the maximum load permitted **was** only 3840 watts.

It is still permissible to use a 15 amp breaker and #14 copper cable to supply 2880 watts or a 20 amp breaker on #12 copper cable supplying 3880 watts but there is no need to do so now. It should be noted that the rules always did permit the cables to carry their full rated current, the trouble was the restriction placed on the breaker supplying the cable. It could only be loaded to 80% of its rating. That has not changed either. What has changed to make this greater loading possible is Rule 62-114(8) which permits higher rated breakers for the #14 and #12 cables when they supply fixed heating loads.

Caution - Do not bundle these cables, maintain separation as described on page 43.

(f)   **Bathroom Heat Lamp** - Rule 62-108(3)

Heat lamps in bathrooms are normally supplied from general use lighting or plug outlet circuits. They may not be supplied from electric heating circuits. The reason is that the heat lamp is not the only heating provided in a bathroom, it is a suplementary heat source for a specific purpose.

(g) **Thermostats**

(i) **Location** - Rule 62-202 requires a thermostat **in each room** where electric heating is installed

Location - The rules do not specify where the thermostat may be located or where it should not be located. It is suggested they should be located at least 36 inches away from the inside wall of a bathtub or shower stall. This suggestion applies to all thermostats. Those separately mounted on the wall and those built into the heater itself.

(ii) **Rating** - Rule 62-118(1) - Those thermostats which are connected directly into the line and control the full load current must have a current rating at least equal to the sum of the current ratings of all the electric heaters they control.

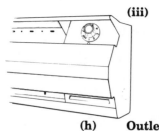

(iii) **Type** - Rule 62-118(2) accepts thermostats whether they are marked with an indicating "off" position or not, Sub-Rule (2) of this rule merely indicates that if a thermostat has a marked "off" position that it must then open all ungrounded conductors of the controlled heating circuit. If the thermostat only indicates a high and a low position with graduated markings between, it need not open all ungrounded conductors of the circuit. Therefore, a single pole thermostat which does not have a marked "off" position would be acceptable on a normal 240 volt heating circuit.

(h) **Outlet Boxes** - Rules 12-506, 12-3002(5)

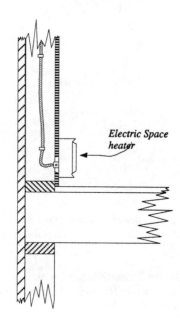

Electric Space heater

Location - An outlet box may be installed behind the heater but that is not the best way to make the connection. This is perhaps the poorest method of connection to use because the boxes must be very accurately located to ensure they will be behind the heater and that they will be properly covered when the heater is installed. Bonding is a problem with this arrangement and finally, access to splices in the box is unsatisfactory. Avoid this method if possible. It is better to run these cables into the thermostat outlet boxes as shown in the circuit illustrations above. Only one cable need enter the heater and this should run in directly without a connection box.

(i) **Cable Protection** - Rule 12-518

Where part of the cable is run exposed to mechanical injury, use a short length of EMT conduit or flexible conduit to protect the cable. Be sure to terminate the conduit in an approved manner so that it is grounded.

(j) **Grounding & Bonding** - Rules 10-400, 10-906

The bare grounding wire (the Code now calls this a bonding wire) in the supply cable must be securely connected to the fixture with the grounding terminal screw in each fixture.

25 ELECTRIC HOT AIR FURNACE - Rule 62-208

Installing an electric hot air furnace is not as difficult as it may appear. By carefully following these basic instructions, you can do it and save.

(a) **Clearances** - Rule 62-208

This rule does not specify any minimum clearance but it does draw attention to two basic clearances required.

(1) **From Combustible Surfaces** - Observe all the clearances specified on the name plate and,

(2) **For Maintenance** - do not install your furnace in a small confined space unless it is designed and marked for installation in an alcove or closet. There must be sufficient clearance to allow removal of panel covers and for maintenance work.

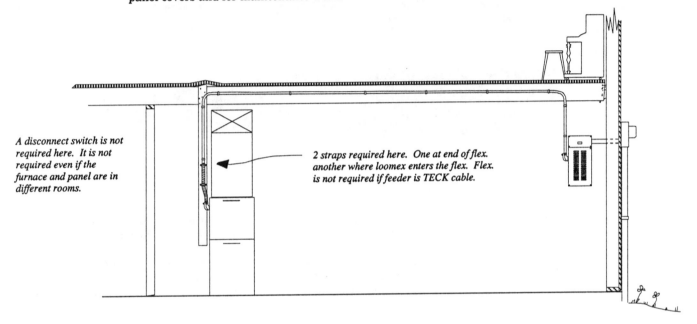

A disconnect switch is not required here. It is not required even if the furnace and panel are in different rooms.

2 straps required here. One at end of flex. another where loomex enters the flex. Flex. is not required if feeder is TECK cable.

(b) **Supply to Furnace** - Rule 62-114(6)&(7)

Below is a greatly simplified table of sizes. Use this table to determine the rating of the furnace supply cable and circuit breakers.

Note - The table also takes into account the current drawn by the fan motor.

Your Furnace Nameplate Rating	Fuse or Circuit Breaker Rating	Size of supply conductor to furnace	
		Using Copper	**Using Aluminum**
5 KW	30 Amp	#10 NMD90 (Loomex Cable)	#8 NMD90 (Loomex Cable)
10 KW	60 Amp	#8 NMD90 (Loomex Cable)	#6 NMD90 (Loomex Cable)
15 KW	100 Amp	#4 NMD90 (Loomex Cable)	#3 NMD90 (Loomex Cable)
20 KW	125 Amp	#3 NMD90 (Loomex Cable)	#1 NMD90 (Loomex Cable)
25 KW	150 Amp	#2 NMD90 (Loomex Cable)	#0 NMD90 (Loomex Cable)
30 KW	175 Amp	#1 NMD90 (Loomex Cable)	#00 NMD90 (Loomex Cable

The above Table is based on Chromalox SPEC. sheet CCSL 266.4. dated Oct. 18/82

(c) **Furnace Disconnect switch**

A disconnect switch is not normally required at the furnace. Check this with your Inspector.

(d)   **Wiring Method**

NMD-9O cable. This cable may be used if the size you need is available. All the usual rules for running loomex cable apply to this cable, i.e. strap it, protect it, and terminate it in approved cable connectors.

Teck Cable (copper) may also be used to supply the furnace. This is a tougher cable and it costs a lot more but it may not need any additional protection from mechanical injury.

Dry type cable connectors may be used indoors.

Note   Peel off only the PVC outer sheath, and only at the connector, so that it (the dry type connector) is clamped directly onto the metal armour.

(e)   **Mechanical Protection Required**

Teck Cable has an aluminum armour and can take a little abuse, however, if there is any doubt, provide protection for the cable especially those sections which are run on the surface below the 5 ft. level.

NMD7 or NMD-90 - Loomex Cable - has no armour and therefore must always be protected from mechanical damage where it is run on the surface below the 5 ft. level. Rule 12-518. Use a flex conduit, sized large enough to allow the cable to move freely, for those sections where mechanical protection is needed.

Note - This flex conduit must then terminate in a flex connector at the furnace.

Consider using an angle connector at the furnace because this cable may not be bent sharply without damaging it.

(f)   **Thermostat Control Wiring**

Table 19 in the code lists only one cable acceptable for low voltage thermostat control work. That is type LVT cable. Truth is, Table 19 does list a type ELC cable for class 2 wiring but Rule 16-210(3) says it may not be used for heating control circuits. So we are stuck with LVT cable only.

Strapping - Whatever staples or straps you use to support this small cable make sure they do not damage the sheath. Staples should be driven in only till they contact the cable. When a staple is driven in too deeply and short circuits the conductors, it turns the furnace on and there is nothing to shut it off, automatically, except the high limit safety device inside the furnace. The thermostat cable is important - install it carefully.

(g)   **Grounding -** Rules l0-400, 10-600

Make certain the grounding conductor is properly connected, with approved lugs, at each end. Look for a separate grounding lug in the furnace connection box and in the service panel.

26 DRIVEWAY LIGHTING

(a) Conduit System

Conduit is very seldom used for this work. It is difficult to properly terminate the conduit at either end. Dont forget, a grounding conductor must be drawn into the conduit along with the circuit conductors to ground the light standard.

Polyethylene pipe manufactured to CSA standard B137.1-1970 may be used for the underground portion only. It need not be CSA certified for electrical use. The riser at the house must be rigid P.V.C. or rigid metal conduit.

(b) Direct Burial - Rule 12-012 & Ontario Bulletin 12-2-8

Depth of Burial - Type NMWU or NMW10 loomex cable, size #14, may be used as shown in the illustration. It must be buried to a depth of 600 mm. (23.6 in.) or if it passes under a roadway or driveway, to a depth of 900 nun. (35.4 in.).

This depth may be reduced by 6 in. (150 mm.) if the cable is protected with treated planking as shown in the illustration.

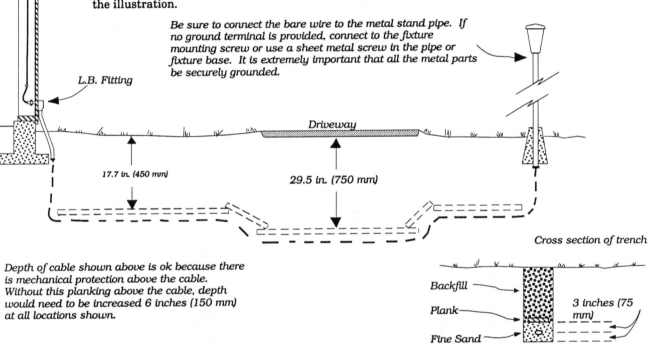

Be sure to connect the bare wire to the metal stand pipe. If no ground terminal is provided, connect to the fixture mounting screw or use a sheet metal screw in the pipe or fixture base. It is extremely important that all the metal parts be securely grounded.

L.B. Fitting

Driveway

17.7 in. (450 mm)

29.5 in. (750 mm)

Cross section of trench

Depth of cable shown above is ok because there is mechanical protection above the cable. Without this planking above the cable, depth would need to be increased 6 inches (150 mm) at all locations shown.

Backfill

Plank

Fine Sand

3 inches (75 mm)

1 - Sand or Earth - may be used but note the stuff is to be screened with 6 mm. (1/4 in.) screen. Cables should lie on 75 mm. (3 in.) thick bed of this screened sand or earth and have a further 75 mm. (3 in.) thick blanket of screened sand placed on top of the cables.

Treated Planking - 2 in. untreated cedar planking is accepted in most cases. The rule says a 38 mm. (1½ in.) treated plank is required. Check with your Inspector before you use untreated planks.

3 - Section of Conduit - This is to protect the cable from damage.

Rule 12-012(5) requires that this conduit terminate in a vertical position approximately 300 mm. (11.8 in.) above the trench floor. The cable must continue on downward, as shown, to allow movement during frost heaving.

Size of Conduit - is not critical. It should be large enough so that cable can be drawn in easily and without damage to the insulation.

Type of Conduit - may be rigid metal conduit or P.V.C. but may not be EMT, Rule 12-1402, 22-500.

(c) Grounding & Bonding - Rules 10-400, 10-906

It is very important that the ground wire be securely connected to the lamp standard.

27 TEMPORARY CONNECTION TO PERMANENT BUILDING - Ontario Bulletin 2-2-3

This is often referred to as TPB. It is really a permanent type of service connection but it is on a temporary basis until the wiring is entirely completed. Many of the building contractors ask for this connection to speed up the finishing work. Check with your Inspector before attempting to prepare the installation for such a connection.

(a)   **Requirements for Connection**

To obtain this connection, in most districts the following is required.

Service - Service must be entirely complete, including;

> — Meter backing
> — Meter blank cover
> — Weather seal below meter
> — Dux seal in last fitting, where required. See page 24 and 29.
> — Service grounding
> — Waste pipe bonding (Note: both septic and soap systems must be bonded if they are metallic.
> — Panel covers must be installed.

Permits - must be complete, that is, they must cover the entire installation, including such appliances as range, water heater, dryer, furnace wiring etc. even though these appliances are not yet installed at the time the temporary connection is required.

One branch circuit must be completed by installing all fixtures, plates etc. but leave all other circuits disconnected unless these too are entirely completed. Do not energize any circuit which runs through an outdoor outlet box unless it is entirely complete with fixtures, fittings, plates, covers etc. It is better to choose a circuit which has all its outlets facing indoors.

In addition, most localities require a temporary connection permit (or TCP) before the Inspector can authorize the power utility to connect the service.

> **Building** - In some localities it is necessary to close the building for this connection. That is, it must have doors and windows installed before the service can be authorized for connection. Check with your local Inspector.

> **Identify House** - The house number should be posted in a conspicuous place on the house. This is to help the Inspector and the power utility connection crew to identify the correct house quickly.

> **Completion** - When the installation is entirely complete, notify the Inspection department immediately. The Inspector requires written notification (this may be a permit stub properly signed and dated or a special form may need to be filled in). Check with your Inspector. If the temporary permit expires before the installation is completed, you will need to renew this connection permit.

Special Cases. Check with your Inspector first.
For Underground Services Only

> It is not always necessary to frame the whole building, i.e. put on the roof and outside sheathing to get a temporary to permanent connection, as described above. Often it is possible to have the building contractor erect only a small portion of the permanent wall where the permanent service (meter base and service panel) is to be located. The permanent meter base and service panel along with one or more plug outlets may then be installed on this partial wall. A temporary wooden enclosure similar to that described for temporary construction service on page 109, must be built to protect this equipment from the weather and small prying and inquisitive fingers from getting hurt.

> **Service Grounding** - Permanent service grounding should be installed. This usually consists of 2 - 3 m(118 in.) rods 3 m (118 in.) apart. Don't forget to connect the water piping system when it is installed. See page 40 for more detailed instructions.

28 TEMPORARY POLE SERVICES - Section 76 in the Code

(a)   **Minimum Service Size**

Wire size...#10
Conduit size ....3/4 inch
Circuit breaker or Fuse ratings ....15, 20 or 30 amp.

Circuit Breaker or Fuse Rating - Rule 28-200 - Circuit breakers or fuses may be 20 amp. or for the larger power saw loads, may be as high as 30 amp. even though the receptacle is rated for only 15 amp.

(b)   **Pole Requirements** - Guide only - check with your Inspector.

    (i)   **Solid** - It must be a solid timber. Laminated timber is not acceptable.

    (ii)   **Size** - Minimum size is 150 mm by 150 mm (6 in. by 6 in.) timber, or if a round pole is used, it must have a 150 mm. (6 in.) diameter top.

    (iii)   **Length** - Minimum length is 5 m (16.4 ft.). This provides for 1.2 m. (47.2 in.) in the ground and a minimum 4.3 m. (14. 1 ft.) above ground. This is acceptable only where lines are short and the Hydro pole is on your side of the road.

         Lines crossing public roadway must be 6 m (19.7 ft.) high.
         Lines Crossing areas accessible to pedestrians only must be a minimum 4.5 m (14.75 ft.) high.

    (iv)   **Gain** - Two shallow saw cuts approximately 2 inches apart, with wood chip between removed, marking the pole 3.6 m. (142in.) from butt end.

    (v)   **Bracing** - as required to offset the pull of the Hydro lines. This should be done with 50 mm x 100 mm. (4 in. x 2 in.) lumber attached as high as possible and should make an angle approximately 30 degrees with the pole.

(c)   **Meter Base** - Ontario Building Code

    (i)   **Height**- maximum 1850 mm (72 in.), minimum 1650 mm (64 in.) grade level to center of meter.

    (ii)   **Blank Cover** - Wood blank cover may be required on the meter base check with your Inspector.

    (iii)   **Connection** - Connections are as shown on page 108.

(d)   **Service Equipment**

    (i)   **Circuit Breaker Type** - This is the preferred type of equipment for temporary construction services.

         Note   Rule 14-302 requires a double pole breaker with single handle be used. Two breakers with a tie-bar on the handles is not acceptable for this purpose because this is a service switch.

Fuse Type — Rule 14-204 - Non-interchangeable type fuses only may be used . This means that fuse adapters must be installed in the standard fuse sockets in the switch and panel. These adapters are available in different sizes or current ratings. Once a particular adapter has been installed, say a 20 amp. size, only a 20 amp. fuse will fit this socket.

Please note: In some districts only circuit breaker type equipment is permitted. You should check with your local Inspector before proceeding.

REJECTION WASHERS

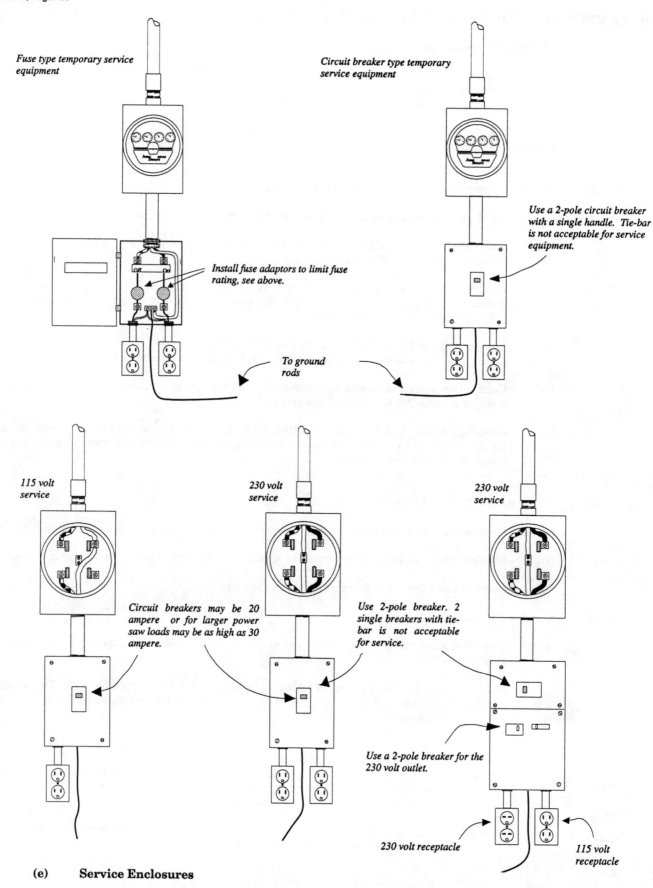

Fuse type temporary service equipment

Circuit breaker type temporary service equipment

Use a 2-pole circuit breaker with a single handle. Tie-bar is not acceptable for service equipment.

Install fuse adaptors to limit fuse rating, see above.

To ground rods

115 volt service

230 volt service

230 volt service

Circuit breakers may be 20 ampere or for larger power saw loads may be as high as 30 ampere.

Use 2-pole breaker. 2 single breakers with tie-bar is not acceptable for service.

Use a 2-pole breaker for the 230 volt outlet.

230 volt receptacle

115 volt receptacle

(e)   Service Enclosures

Standard Non-Weatherproof Type - Non-weatherproof service equipment may be used if it is installed in a solidly constructed **weatherproof box.** This box may be of lumber or plywood not less than 20 mm (3/4 in.) in thickness with hinged door and lock. If practicable, the door should be hinged at the top.

Standard Equipment in Wooden Enclosures

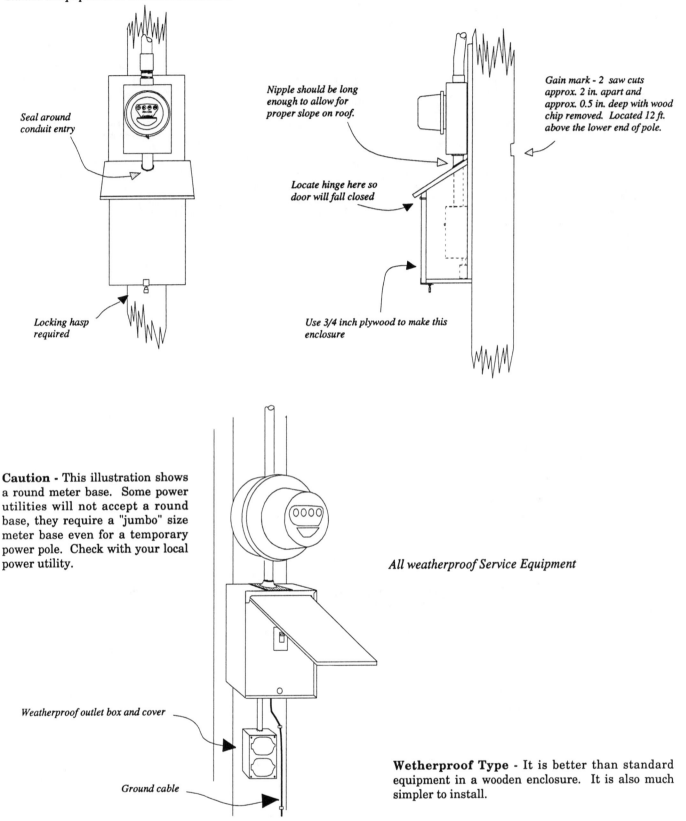

Seal around
conduit entry

Locking hasp
required

Nipple should be long
enough to allow for
proper slope on roof.

Locate hinge here so
door will fall closed

Use 3/4 inch plywood to make this
enclosure

Gain mark - 2 saw cuts
approx. 2 in. apart and
approx. 0.5 in. deep with wood
chip removed. Located 12 ft.
above the lower end of pole.

Caution - This illustration shows
a round meter base. Some power
utilities will not accept a round
base, they require a "jumbo" size
meter base even for a temporary
power pole. Check with your local
power utility.

All weatherproof Service Equipment

Weatherproof outlet box and cover

Ground cable

Wetherproof Type - It is better than standard
equipment in a wooden enclosure. It is also much
simpler to install.

Note - Use an FS outlet box (this is a cast iron box suitable for use outdoors) and spring-loaded, gasketed cover plate
over the receptacle as shown. See also Rule 26-706.

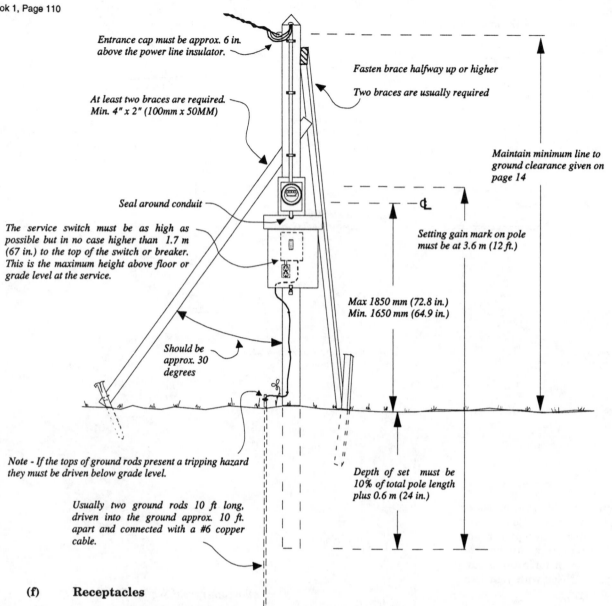

Entrance cap must be approx. 6 in. above the power line insulator.

At least two braces are required. Min. 4" x 2" (100mm x 50MM)

Fasten brace halfway up or higher

Two braces are usually required

Maintain minimum line to ground clearance given on page 14

Seal around conduit

The service switch must be as high as possible but in no case higher than 1.7 m (67 in.) to the top of the switch or breaker. This is the maximum height above floor or grade level at the service.

Setting gain mark on pole must be at 3.6 m (12 ft.)

Max 1850 mm (72.8 in.)
Min. 1650 mm (64.9 in.)

Should be approx. 30 degrees

Note - If the tops of ground rods present a tripping hazard they must be driven below grade level.

Depth of set must be 10% of total pole length plus 0.6 m (24 in.)

Usually two ground rods 10 ft long, driven into the ground approx. 10 ft. apart and connected with a #6 copper cable.

(f) Receptacles

(i) **120 Volt Receptacles** - A 2 wire, 120 volt service would have a single pole circuit. It could serve one 120 volt plug receptacle.

(ii) **A 3-Wire, 120/240 Volt Service** would have a 2-pole sircuit breaker. It could serve two 120 volt plug receptacles or one 240 volt receptacle without requiring a branch circuit panel.

(iii) **240 Volt Receptacles** - Whenever a 240 volt plug outlet is required, in addition to a 120 volt outlet, you must install a branch circuit panel. A combination circuit breaker panel is the best way out of this awkward requirement.

(g) Grounding

(i) **Cable** - must be of copper. Where the service is 100 amp. or less the grounding conductor may be #4 bare or even #6 bare copper if not subject to mechanical damage. See also page 39 for more detailed information.

(ii) **Ground Rods** - Minimum size required.
Copper rods......................$\frac{1}{2}$ inch by 3 m. (118 in.)
Solid iron rod ....$\frac{5}{8}$ inch by 3 m. (118 in.)

Note - Use solid rods - pipe is not approved any longer. Rule 10-702.

(iii) **How Many Rods?** - In most localities 2 rods driven 3 m (approx. 10 ft.) apart, are required. In others you may need more to do the job adequately.

29 PRIVATE GARAGE & FARM BUILDINGS

(a) **Overhead Supply Lines**

(i) **Insulation of Wire** - Rule 12-302 - Conductors must have weather proof insulation. Triplex cable is normally used today. Rule 12-318.

(ii) **Elevation of Wire** - Rule 12-304 - Overhead lines must be out of reach, out of harms' way and according to this rule, be at least 4.5 m. (14.8 ft.) above ground. This is obviously not obtainable or necessary in every case, however, before proceeding with anything less than 4.5 m. (14.8 ft.) you should check with your Inspector.

(iii) **Roof Crossing** - Rule 12-312 - Conductors shall not be carried over buildings without special permission and work shall not begin until the plans and specifications for the work are approved by the Inspection department. You might as well know it now, special permission to cross a roof is not easily obtainable.

(iv) **Triplex Type Cable** - Rule 12-318 - A single 15 amp. branch circuit may be run to a building as shown below. The simplist method is to use a triplex cable. This consists of two insulated conductors wraped around a bare messenger cable. In some cases the bare messenger, in triplex cable, may be used as the bonding conductor between the two buildings.

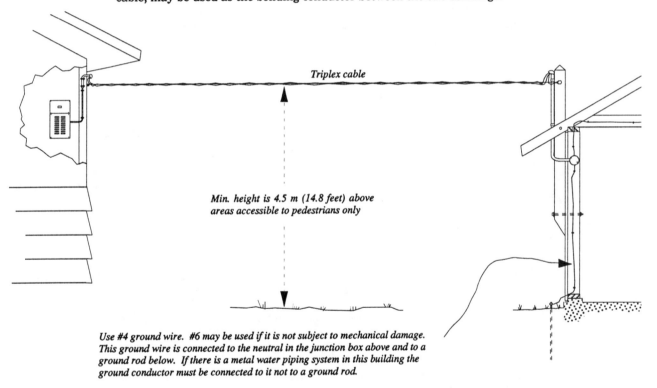

Triplex cable

Min. height is 4.5 m (14.8 feet) above areas accessible to pedestrians only

Use #4 ground wire. #6 may be used if it is not subject to mechanical damage. This ground wire is connected to the neutral in the junction box above and to a ground rod below. If there is a metal water piping system in this building the ground conductor must be connected to it not to a ground rod.

Two 15 amp. circuits may be run overhead as shown but if more circuits are needed a sub-panel is required in the out-buildings.

(v) **Grounding** - Rule 10-208 - The neutral must be bonded to ground at each building. If this building **does not house livestock** the bare messenger in the triplex cable may be used to ground the circuits. If the building is used for livestock the circuits must be grounded at the building as shown in the illustration. This grounding conductor is usually #4 bare copper. A #6 conductor may be used if it is not subject to mechanical damage, Rule 10-806. The connection may be made in the first junction box. Do not use solder for this connection. Normally, only one ground rod is required for each sub-service.

(b)   **Underground Supply Lines - Rule 12-012**

Cable Type —Use NMWU - Table 19 (This is the old NMW10)

Cable Size — Use #14 but for long runs, above 15 m. (50 ft.), #12 is recommended.

Cable Depth - without a protecting plank above the cable

     600 mm (23.6 in.) under pedestrian only area.
     900 mm (35.4 in.) under vehicular traffIc areas.

With a protecting plank above the cable as shown below.

     450mm (17.7 in) under pedestrian only area.
     750 mm (29.5 in.) under vehicular traffic areas.

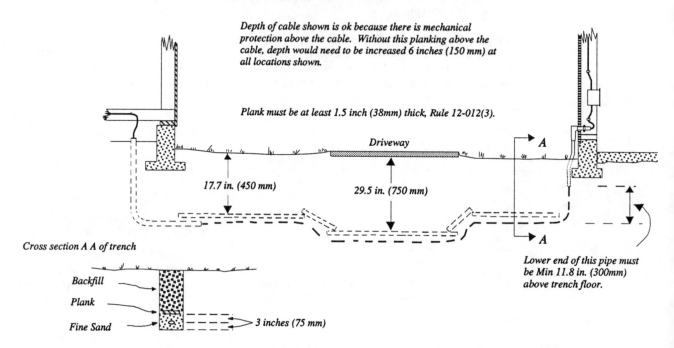

Depth of cable shown is ok because there is mechanical protection above the cable. Without this planking above the cable, depth would need to be increased 6 inches (150 mm) at all locations shown.

Plank must be at least 1.5 inch (38mm) thick, Rule 12-012(3).

Driveway

A

17.7 in. (450 mm)

29.5 in. (750 mm)

A

Cross section A A of trench

Lower end of this pipe must be Min 11.8 in. (300mm) above trench floor.

Backfill
Plank
Fine Sand

3 inches (75 mm)

Cable Protection -   Conduit should extend into the trench as shown and cables underground should be protected as follows:

     75 mm (3 in.) of sand below cable
     75 mm (3 in.) of sand above cable.
     38 mm (1½ in.) plank above the sand. (Not nominal but actual thickness).

Polyethylene pipe, manufactured to CSA Standard B137.1, may be used for the underground portion only. The riser at each end must be either rigid PVC or rigid metal conduit, not EMT.

(c)   **SUB-PANELS** - in a garage or similar building.

Where the load in the garage or other separate building requires more than two 15 amp. circuits, or where you wish to provide for future load additions, a sub-panel may be installed.

Sizes   Where plug outlets are used to supply power tools in a private garage or workshop it is recognized that there will normally be only one man working these tools at any one time. Therefore, even if you have a number of large power tools it is unlikely that more than one tool would be used at any one time. For this reason, when we are calculating the feeder size for this panel, we need to concern ourselves with only one machine, the one with the largest electrical load.

Wood and Metal Lathe

Drill Press

Power Saw

#10 Sub-Feeder - If there is no fixed electrical heating load and no electric welder load a #10 NMD-90 (copper) cable may be sufficient. This cable could supply 30 ampere load at 240 volts.

It could supply a 1½ HP power tool at 240 volts, one circuit for lighting and one for plug outlets. The panel should be large enough for present loads plus some reserve capaciry for future load additions.

LARGER LOADS - such as welders or fixed electric heaters would require more power and therefore a larger cable. A #8 NMWU (copper) cable will supply 40 armperes at 240 volts. This is large enough to supply a small private garage type welder, one circuit for lighting and one or more circuits for plug outlets. If the full load current of the welder was say 20 amps, then the circuit breaker supplying the 40 ampere NMWU sub-feeder cable could be as high as 60 amperes. Obviously, the load may not be greater than 40 amperes on this 40 ampere cable but because the biggest load is either a welder load or a motor load the code permits the ampere rating of the main supply breaker to be greater than the cable rating.

WOODWORKING SHOP - This 40 ampere NMWU copper feeder cable supplied with a 60 ampere circuit breaker in the main panel could serve 3 HP 240 volt motor. There would be sufficient capacity left to supply garage lighting and a number of branch circuits for plug outlets.

THE BIG ONE - A #6 NMWU (copper) feeder to a garage or workshop is better but it may be more than is necessary in most cases. If you have an unusually large load, or the length of this feeder to the 2nd

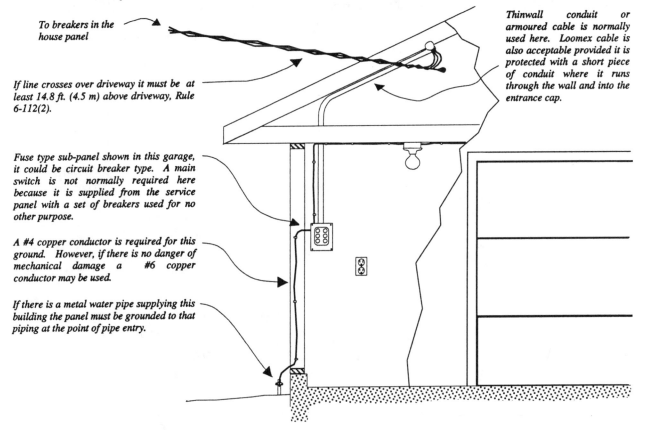

To breakers in the house panel

If line crosses over driveway it must be at least 14.8 ft. (4.5 m) above driveway, Rule 6-112(2).

Thinwall conduit or armoured cable is normally used here. Loomex cable is also acceptable provided it is protected with a short piece of conduit where it runs through the wall and into the entrance cap.

Fuse type sub-panel shown in this garage, it could be circuit breaker type. A main switch is not normally required here because it is supplied from the service panel with a set of breakers used for no other purpose.

A #4 copper conductor is required for this ground. However, if there is no danger of mechanical damage a #6 copper conductor may be used.

If there is a metal water pipe supplying this building the panel must be grounded to that piping at the point of pipe entry.

building is say 80 ft or more, you should consult your local Electrical Inspector for advice.

Grounding - Rule 10-208 - says the neutral must be grounded at each building which houses livestock but for all other buildings Subrule (b) suggests that grounding can be brought in with the sub-feeder conductors. If this sub-feeder is run underground direct from the service panel it is easy to bring a grounding conductor with the feeder but when the supply feeder is run overhead it may be much more difficult. While all that is possible there is yet another possible problem with that arrangement. If there is a metal piping system entering this building it must be bonded to this incoming grounding conductor. This is too complicated. The normal arrangement is as shown.

This grounding conductor is usually a #4 bare. A #6 grounding conductor may be used if it is not subject to mechanical damage, Rule 10-806. This grounding conductor must be connected to tne neutral bus in the first switch or branch circuit panel in the second building. If the neutral in that switch or branch circuit panel is not already bonded to the enclosure you will need to bond it as shown for services on page 39.

Ground Rods - Normally, only one 3 m (10 ft.) ground rod is required for each sub-service panel in a separate building.

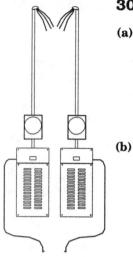

30 DUPLEX DWELLINGS

(a)   **Separate Service Conductors** - Rule 6-104 - In duplex dwellings, where it is not possible to locate common service equipment, each suite may be served with separate service equipment as shown. The meter base, service panel and all the instructions for grounding would be exactly the same as that given in the appropriate sections in this book for a single family residence.

> **Note**   The two entrance caps must located close together so that only one Hydro service drop is required.

(b)   **Common Service Conductors** - Rule 6-200(2) - This rule permits a two-gang meter base to be used in a duplex dwelling without a main switch ahead of it

> **Note** - The main service conductors must be large enough to carry 100% of the calculated load in the heavest loaded suite plus 65% of the calculated load in the other suite. See sample calculation

(i)   **Calculation** - If both halves of the duplex are to be served from a common service the size of these conductors should be as follows:

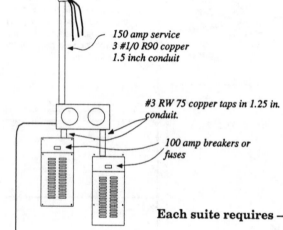

150 amp service
3 #1/0 R90 copper
1.5 inch conduit

#3 RW 75 copper taps in 1.25 in. conduit.

100 amp breakers or fuses

Sample Calculation - Identical load in each suite.

Basic Load 102.2 m² (1100 sq.ft.) floor area

First 90 m² (968.4 sq.ft.)	5000 watts
Next 12.2 m² (131.6 sq.ft.)	1000 watts
10kw.range	6000 watts
4 kw. dryer	1000 watts
3 kw. water heater	750 watts
Electric heating (assume 7kw.)	7000 watts
Total	20750 watts

$$\frac{20750}{240} = 86.5 \text{ amps}$$

Each suite requires — 100 amp. service panel with 100 amp. sub-breaker.
#3 RW75 copper sub-service wire in 1¼ in. sub-service conduit.

Main Service - To determine main service size add the loads as follows:

Largest suite load is	100% of	86.5 amps	= 86.5 amps
Add second suite	65% of	86.5 amps	= 56.2 amps
	Calculated service service size		= 142.7 amps

(ii)   **Metering and Service Conductors** - Normally the suite service panels can be installed within a few feet of the meter base, as shown above. Rule 6-206(1)(e) limits service conductor length inside the building to "as close as practicable" which means just through the wall plus a tiny little bit. On the outside of the building these runs may be as long as needed except that they may not be more than 7.5 m (24.6 ft.), Rule 14-100(c).

(iii)   **Service Equipment** - The service equipment in each suite normally consists of a combination service panel as shown.

(iv)   **Service Grounding** - In this case the service grounding conductor must run into the duplex meter base. These meter bases usually have provision for the grounding conductor connection and for bonding the netral to the meter enclosure. See also under "Grounding" on Page 39.

Size of Grounding Conductor - may be determined from the Table, page 6 in this book.

> **Note**   The size of this grounding conductor is related to the current carrying capacity of the main service conductor.

31 REWIRING AN EXISTING HOUSE

(a)   Electrical Service Panel

Old Panel - If you are adding load to an existing panel in an old house but have no more breaker or fuse spaces you will need to install an additional sub panel and feed it from breakers in the existing panel. Before you do this, make sure the service equipment is adequate to supply the new load. Use the table on page 6 to determine size of service needed for all the load. If there is sufficient spare capacity, reroute two of the existing circuits to a new panel located nearby. This will free up space for two breakers for a sub-feeder to supply the new panel.

Change Service panel - This will cost a bit more but it may be the only solution if the old stuff is too small. Don't forget that if you change the panel you must upgrade the whole service, not the branch circuit wiring but the service. The reason for this is the new service may require larger conductors, conduit, meter base and grounding. It may also be necessary to find a more acceptable location for the new panel, meter base or service entrance cap. It may also be that all of these items are satisfactory and all that is needed is to replace the old panel with a larger one.

All this may sound expensive and indeed may be but don't forget this is the most important part of your installation. This is what protects all the circuits in the house and this is where that enormous hydro electric capacity is limited to a safe value. To compromise here is to compromise with minimum safety.

Caution - Even if it looks harmless, Don't take chances. Before any work is begun you should make sure that that part of the system you will be working on is, in fact, not energized.

If you are changing your service equipment have Hydro crew disconnect your service before you begin work.

Where a service has been changed the Inspector will usually check for:

- **Entrance cap** - Re: height and location, page 14.

- **Size of Service Conduit & Conductors** - Bulletins 2-3-3 and 12-15-6 - This may have been large enough at one time but may be too small now. See under "Service Size", page 5.

- **Meter Location etc.** - Accessibility and distance from the front of the house. page 30.

- **New Panel** - Must comply with new Code re: ampacity, number of circuit positions and number of 2-pole positions available for 240 volt and 3 wire circuits, page 35.

- **Grounding** - Size of conductor, condition of run, connected to old abandoned water service? Must comply with current regulations, see page 39.

- **Bonding** - Rule 10-406(3) - Older houses may have metal waste pipe system which may not have been bonded. In that case the code requires a bonding connection between the metal waste piping and the nearest cold water pipe.

(b)   Branch Circuit Wiring

The Inspector will not normally require the existing finished areas to be rewired except as may be necessary for minimum safety. He will, however, require upgrading of branch circuit wiring where the walls have been opened to make other structural revisions or additions to the house. He has the knowledge and the experience to guide him in his assessment of your installation. He will ask for minimum changes to the existing wiring to make your home reasonably safe. The fact is - it's your safety he is concerned about and you want him to be concerned about that.

Additions To House. - All new wiring in the new addition must comply with current regulations. The information on branch circuit wiring given in this book does apply to the additional wiring.

Caution - Before entering the service panel make sure the main switch or breaker is in the "off" position.

Electrical Permit - make sure your electrical permit is adequate to cover the work done.

32 MOBILE HOMES

(a) **C.S.A. Marking** - Rule 2-020

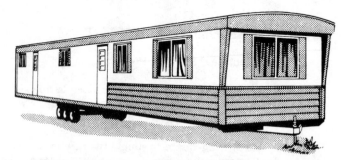

The mobile unit must have a C.S.A. label or a label applied by one of the other certification agencies noted on page 2. If your unit does not have such a label you are in deep trouble. Sometimes it is lost in a move or for one reason or another it is gone. This approval label is one of the first things the Inspector will look for. If it is not there he cannot authorize a service connection. Your local inspection office can advise you on the proceedure for replacing a lost certification label.

(b) **Electrical Permit** - Rule 2-004

An electrical permit must be obtained before any electrical work is done.

Note   If this mobile unit is on private property and you are personally going to occupy it, you are usually permitted to obtain a permit and do the work; or

If this mobile unit is to be occupied by someone in your immediate family, you are usually permitted to obtain an electrical permit and do the work; but

If this unit is for rental purposes to someone other than the immediate family then the wiring must be installed by an electrical contractor; and

If this unit is to be connected in a mobile home park the work must be done by a certified electrical contractor. The owner would not be permitted to do this work nor to take out a permit.

(c) **Service Size** - Rule 72-102(1)

Mobile homes are treated the same as a single family dwelling as far as service ampacity is concerned. See the table on page 9 for quick sizing of service equipment.

Rule 8-200, Floor area - **less** than 80m² (861 sq.ft.)
The normal load will be something like this.

Basic	5000 watts
Range - (12 kw. range)	6000 watts
Water heater - (3 kw tank)	750 watts
Dryer — (4 kw. unit)	1000 watts
Quick Steam Tap — (500watts)	0 watts
Furnace — (gas or oil)	0 watts
Freezer	0 watts
Dishwasher	0 watts
Fridge	0 watts
Total	12750 watts

12750 divided by 240volts = 53.1 amps.

The code requires a 60 amp. service for this load.

Note   It really does not matter if the range, dryer, water heater and furnace are all gas - the fact is, the minimum service size is still 60 amp. for any size floor area up to 80m² .

Larger Floor Area - Where the floor area **is** 80m² (861 sq. ft.) or larger the minimum size service acceptable by Code is **100 amps**.

Unfortunately, the rules make no distinction between a single family house and a mobile home. The same rules used to determine the size of a service in a house are also applied in determining the size of service in a mobile home.

We all know that the calculated load in this unit is much less than 100 amp. The extra service ampacity is usually intended for future load changes. In single family houses there is always the possibility of finishing a basement, adding a sauna or large power vacuum etc. In a mobile home such changes are difficult, very costly and rare. For this reason check with your local Inspector if he will accept a 60 amp. service for this teensy bit larger floor area.

(A) SINGLE UNIT IN MOBILE PARK

(a)   Connection Box in Mobile Home Park

A typical connection box in each lot in a mobile home park.

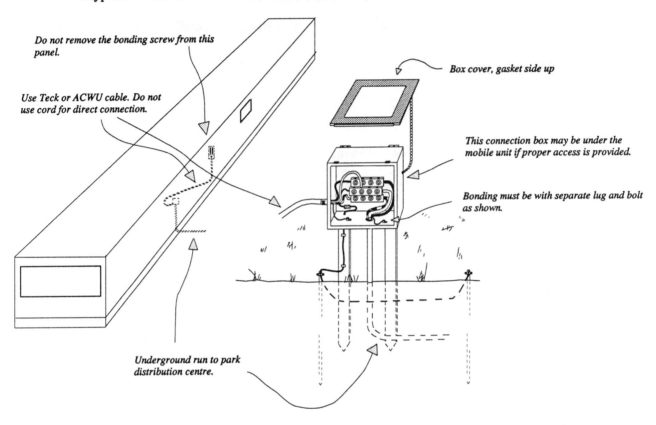

Do not remove the bonding screw from this panel.

Use Teck or ACWU cable. Do not use cord for direct connection.

Box cover, gasket side up

This connection box may be under the mobile unit if proper access is provided.

Bonding must be with separate lug and bolt as shown.

Underground run to park distribution centre.

Connection to this box can only be made by a certified electrical contractor. Neither the owner of the mobile unit nor the mobile park operator (or owner) may make this connection unless they have a valid Certificate of Qualification to do this work.

An electrical permit must be obtained before connection is made.

Cable Connector - The cable from the mobile unit must connect to the box with a connector approved for the particular cable you are using.

Conductor Terminations in Box - Watch this carefully - most of the problems are right here.

-       **Don't skin back too much insulation** - only enough to make good connections.

-       **If the box does not have** a terminal strip as shown - use only approved connectors to make the splices. If aluminum conductor is used, this is extremely important. In that case use only connector lugs marked approved for aluminum.

-       **Bonding** - Terminate the grounding conductor in an approved lug which is separately bolted to the box. Don't try to use a bolt or screw already used for some other purpose.

-       Clean the Surface under the bonding lug to ensure a good electrical contact.

(B) SINGLE UNIT ON PRIVATE PROPERTY

(a) Basic Requirements

(i) Permission to Place Mobile Home

More and more of our freedoms are being taken away from us. Before you move a mobile unit onto your own property, it is best to find out first if that would be permitted by local by-law. In some cases you will be required to produce some kind of proof of approval from the local authorities before the Electrical Inspection department will issue an electrical permit for connection.

(ii) Consult Hydro — Rules 6-206(1) & 6-114

Before any work is done the power utility should be consulted to determine which pole service will be from.

(b) Connection Methods

There are several connection methods you may choose from - each has its problems and pitfalls.

A   DIP SERVICE

As shown below, only the meter is on the pole. The service conductors dip underground to a point under the panel where they rise and enter the service box in a mobile unit. This means the service conductors do not have fuse protection until they enter the panel. It means there is no quick easy shut-off, except by removal of the meter when the moving truck arrives.

Note - Meter only on service pole.

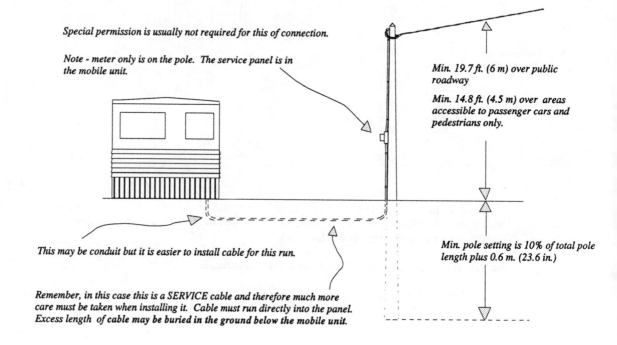

Special permission is usually not required for this of connection.

Note - meter only is on the pole. The service panel is in the mobile unit.

Min. 19.7 ft. (6 m) over public roadway

Min. 14.8 ft. (4.5 m) over areas accessible to passenger cars and pedestrians only.

This may be conduit but it is easier to install cable for this run.

Min. pole setting is 10% of total pole length plus 0.6 m. (23.6 in.)

Remember, in this case this is a SERVICE cable and therefore much more care must be taken when installing it. Cable must run directly into the panel. Excess length of cable may be buried in the ground below the mobile unit.

B **Mast on Mobile Unit**

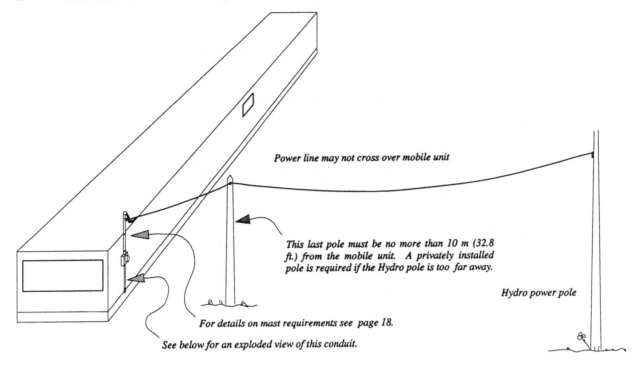

Power line may not cross over mobile unit

This last pole must be no more than 10 m (32.8 ft.) from the mobile unit. A privately installed pole is required if the Hydro pole is too far away.

Hydro power pole

For details on mast requirements see page 18.

See below for an exploded view of this conduit.

For mast support details see page 18.

Note - It is often difficult to provide adequate mast support on a mobile unit. For this reason the Inspection authorty requires that the last pole in the power line to the unit **must be no more than** 10 m (32.8 ft.) from the mobile unit.

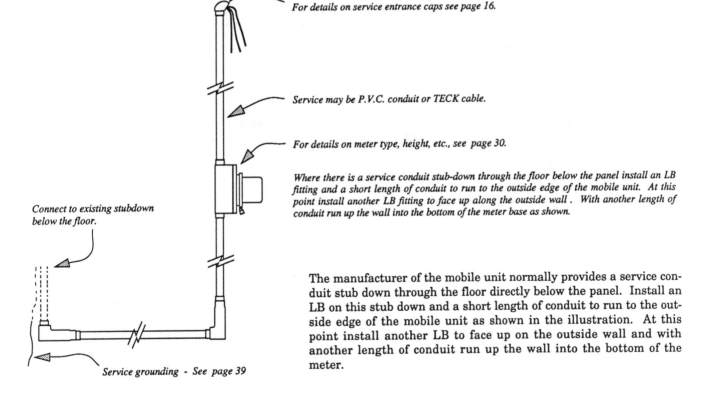

For details on service entrance caps see page 16.

Service may be P.V.C. conduit or TECK cable.

For details on meter type, height, etc., see page 30.

Where there is a service conduit stub-down through the floor below the panel install an LB fitting and a short length of conduit to run to the outside edge of the mobile unit. At this point install another LB fitting to face up along the outside wall . With another length of conduit run up the wall into the bottom of the meter base as shown.

Connect to existing stubdown below the floor.

Service grounding - See page 39

The manufacturer of the mobile unit normally provides a service conduit stub down through the floor directly below the panel. Install an LB on this stub down and a short length of conduit to run to the outside edge of the mobile unit as shown in the illustration. At this point install another LB to face up on the outside wall and with another length of conduit run up the wall into the bottom of the meter.

C Pole Service - By SPECIAL PERMISSION ONLY

This is an outdoor service and Rules 6-206(d) & 70-102(2) says special permission is required for such an installation. Special permission for these installations must be obtained before any work is done. Service ampacity is based on Rule 72-102(1) which is the same as for a house of similar size. See the table on page 9.

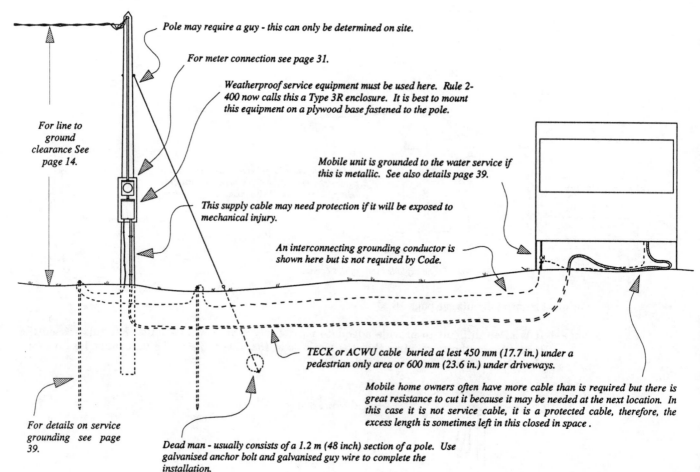

Pole may require a guy - this can only be determined on site.

For meter connection see page 31.

Weatherproof service equipment must be used here. Rule 2-400 now calls this a Type 3R enclosure. It is best to mount this equipment on a plywood base fastened to the pole.

For line to ground clearance See page 14.

Mobile unit is grounded to the water service if this is metallic. See also details page 39.

This supply cable may need protection if it will be exposed to mechanical injury.

An interconnecting grounding conductor is shown here but is not required by Code.

TECK or ACWU cable buried at lest 450 mm (17.7 in.) under a pedestrian only area or 600 mm (23.6 in.) under driveways.

Mobile home owners often have more cable than is required but there is great resistance to cut it because it may be needed at the next location. In this case it is not service cable, it is a protected cable, therefore, the excess length is sometimes left in this closed in space .

For details on service grounding see page 39.

Dead man - usually consists of a 1.2 m (48 inch) section of a pole. Use galvanised anchor bolt and galvanised guy wire to complete the installation.

Grounding -   Use two 3 m. (10 ft.) ground rods and connect to the water service pipe under the unit. See under Grounding page 39 for detailed information.

Cable Types -   ACWU, TECK and Corflex may be used. NMWU may also be used when provided with adequate mechanical protection - not recommended for this application.

Service -   See index for detailed list of requirements.

(c) Pole Location on Property

This is very important — consider the following:

(i) Near Connection Point

The pole should be (not must be) located near the point of service connection on the mobile unit. The length of your existing supply cable will often determine this.

(ii) Length of Hydro Lines Required

Note the Hydro power pole location. Hydro will usually run 30 m (98 ft.) onto your property without additional fee. Beyond this length they may be into your wallet for more money. Long runs, more than 30 m. (98 ft.) may also require additional poles and this is at your expense, usually.

(iii) **Crossings**

- **Crossing a Public Roadway?**

    You may not have a choice. If the Hydro line is on the other side of the road you must cross the road with the line. Clearance required is 6 m. (19.7 ft.) minimum above roadway. It may require a long pole to do this.

- **Crossing a Driveway?**

    Avoid crossing driveways if possible. If you must cross, the minimum line to ground clearance is 4.5 m. (14.75 ft.) above a driveway. This is generally considered the minimum height even if there is no garage or carport, only a driveway.

- **Crossing a roof?**

    No! - Do not cross the roof of any building including the roof of the mobile unit. Locate the pole so as to avoid it.

(d) **Pole Requirements**

Types - Cedar, fir, lodgepole pine are usually acceptable. If you go into the woods to cut down a tree to make a pole, observe the following picky points:

    I - Make sure it is straight and has the correct dimensions - see below under 'length'.

    2 - Remove all branches, leaving a smooth surface.

    3 - Remove all bark.

    4 - Cut a gain mark 3.6 m. (142 in.) from butt, as shown below.

    5 - Apply an effective treatment to both ends to retard rotting.

    Note - It may be possible to use a sawn timber instead of a round pole. It is not recommended and before you try it, check with your Inspector.

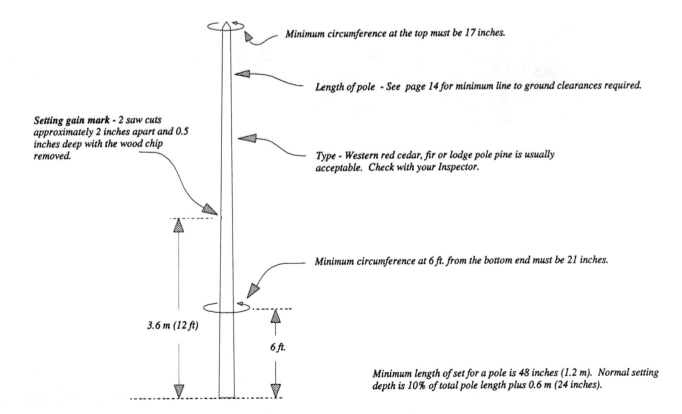

Minimum circumference at the top must be 17 inches.

Length of pole - See page 14 for minimum line to ground clearances required.

Setting gain mark - 2 saw cuts approximately 2 inches apart and 0.5 inches deep with the wood chip removed.

Type - Western red cedar, fir or lodge pole pine is usually acceptable. Check with your Inspector.

Minimum circumference at 6 ft. from the bottom end must be 21 inches.

3.6 m (12 ft)

6 ft.

Minimum length of set for a pole is 48 inches (1.2 m). Normal setting depth is 10% of total pole length plus 0.6 m (24 inches).

Size - Must be minimum Class 6. This means it must be:

1 - Minimum 17 inches circumference at the top of the pole. This is minimum, it may be larger.

2 - Minimum 21 inches circumference 6 ft. from the butt end.

Length - Two factors must be considered.

1 - **Length of pole above ground.** This is simply line to ground clearances given on page 14. Add to this:

2 **Length of pole set into the ground.** In most cases this is 1.2 m. (47.2 in.). Where the pole length above ground must be greater than 6 m. (19.7 ft.) the length of the pole set into the ground must be increased. The normal setting depth is 10% of total pole length plus 0.6 m.

Guy Lines - When Required? This must be determined on site. A very rough thumb rule is as follows:

If the length (of pole above ground) is 4 5m (14.75 ft.) or less and the length of span for the Hydro lines is say 10 m. (59 ft.) a guy is likely not required. If the dimensions are greater than these, it is very likely you will need to guy the pole.

- **Type of Guy Wire** - must be galvinized cable. Do not try to use ordinary wire rope or clothes line cable.

- **Size** - 5/16 in. guy wire would be sufficient in most cases.

- **Dead Man** - This is usually a 4 ft. length of log buried to the same depth as the pole setting. Use a long anchor bolt as shown below.

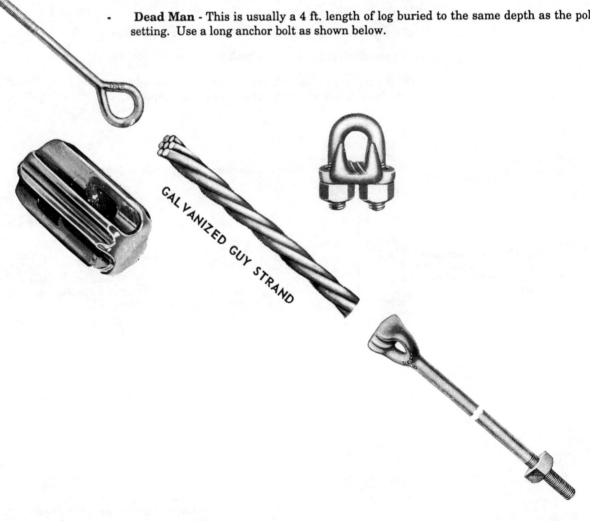

GALVANIZED GUY STRAND

33 PRACTICE EXAMINATION for CERTIFICATION

In the official examination you are not permitted to use your own code book or bulletin book. The examiner will provide unmarked copies of these books to you. He will also supply writing paper for your calculations.

Use all the time allowed for the exam. This is not a test to determine who finishes first but who finishes best.

Don't guess where True or False answers are required. A penalty may be assessed for each incorrect answer. Therefore, you could lose not only the marks for the question but also an equal mark as a penalty.

The theory behind this is that if you knew the correct answer, you would have given it but since you gave an incorrect answer you must have have been guessing. Therefore, if you don't know the correct answer, leave it blank unless you are very good at Russian Roulette.

TYPICAL EXAMINATION QUESTIONS

SECTION 1   TRUE or FALSE

Allow 2 points for each correct answer and subtract 2 points for each incorrect answer.

		True	False
I.	A CSA enclosure 3 may be used indoors where subject to drops of falling liquid due to heavy condensation		
2.	A standard duplex receptacle may be installed in a bathroom provided it is GFCI protected		
3.	The neutral must be bonded to the meter base in every case without exception		
4.	The plug outlet on the bathroom light fixture is acceptable to the code for the bathroom plug		
5.	A diesel fuel pump at a service station is considered less hazardous than a gasoline fuel pump		
6.	The grounding conductor running around a pool must connect to the reinforcing steel of the pool in at least 3 equally spaced points around the pool		
7.	The metal supply conduit to the deck box for an inground pool light fixture may be used to ground the fixture		
8.	The rules permit the G.F.C.l. type receptacle in the bathroom to also serve the outdoor plug outlet		
9.	2 or more loomex cables may be run in contact with each other without affecting their ampacities		
10.	The electrical code specifically states that where two or more smoke alarms are installed they shall be interconnected to sound together		
11.	There is no limit to the length of service conduit if it is run on the outside surface of the building		
12.	A grounding conductor is not required in seal-tight flex provided it is smaller than 1 inch		
I3.	The code does not require the secondary of a 24 volt class II transformer to be grounded when the primary is120 volts and all the secondary wiring is inside the building		
14.	Plug outlets may be located between 60 in.(1.5 m) and 118 in.(3 m) of the inside wall of a swimming pool provided they are protected with a G.F.C.l. of the class A type		
I5.	TWH conductor is rated for 75 C and therefore may be used in conduit for fire alarm system wiring		
16.	Conductors of any size or colour may be painted white and used for neutral conductors except in the case of service conductors		
17.	The disconnect for a hoist must be accessible from the ground or from the floor below the hoist		
18.	The plug outlet for the door opener in a garage may be supplied with a garage lighting circuit		
19.	A telephone jack, if located in a bathroom, must be at least l.2m from a tub or shower		
20.	Electrical metallic tubing must be securely fastened in place within 39.4 inches of each outlet box, junction box. cabinet or fitting		
21.	A 60⁰ C conductor could not carry any current at all in an ambient temperature of 60 C		
22.	Underground wiring to the dispensing pumps at a gas station must conform to code requirements for Class I. Div. 2 hazardous location wiring methods		
23.	If the pump and hot tub are at least 3 m apart G.F.l. protection is not required for the pump		
24.	A circuit supplying a combination heat lamp (maximum rating 300 watts) and exhaust fan may also supply other outlets		
25.	Sectional boxes are now approved for use embedded in concrete		
26.	Set screw type EMT couplings and connectors may be used underground provided the conduit is not buried below permanent moisture level		
27.	An electrical permit is required for all small jobs such as a furnace, range or dryer connection		

28. The wiring for fire alarm systems may not be drawn into conduits used for other circuits. They must be kept entirely separated except at point of supply.. _____   _____

29. It is the responsibility of the Electrical Contractor to ensure that all the electrical equipment he connects has been properly certified by CSA or by one of the other equipment certification agencies .... _____   _____

30. The service conductors for a 60 amp. service in Toronto may be #6TW provided the demand load is not more than 55 amp... _____   _____

31. For a single family dwelling, all outdoor plug receptacles, which are accessible from grade level, must be supplied with a circuit used for no other purpose... _____   _____

SECTION TWO-   FILL IN THE BLANK SPACES.
Each question is worth 2 points.

1. Minimum permissible radius of teck cable bends is ____ the internal diameter of the cable.

2. Junction boxes may be installed in attics, ceiling spaces or crawl spaces provided the head room, vertical clearance, is _____

3. The maximum permissible load on a sign circuit is _____

4. Where low voltage conduit is buried underground it shall be at least _____ below the surface where it is subject to vehicular traffic and _____ where not subject to vehicular traffic.

5. The rated secondary open circuit voltage of a sign transformer shall not exceed _____ volts.

6. Where NMS cable is run through studs, joists or similar wooden members, the outer surface of the cable shall be kept at least _____ from the nailing edges of wooden members or be protected.

7. Where light fixtures are mounted less than _____ above the floor they must be guarded or be flexible.

8. The maximum number of light or plug outlets permitted on a circuit is _____

9. The minimum wiring method for Class I, Division I location is ____

10. The maximum length of a #14 TW tap from a 30 amp. branch circuit for heating is _____

11. Outlet boxes must be set flush with the surface finish except that in the case where the finish is non-combustible type the box may be set back _____

12. The minimum size bonding conductor for a spa is ____

13. The maximum fixed heating load that may be connected to a 30 amp. 240 volt branch circuit breaker, using #12 copper conductors, (in a residence) is _____ watts.

14. The service entrance cap shall be so located that open conductors are above windows etc. or not less than ____ from windows or similar openings.

15. The minimum distance between an electrical meter and a gas meter is _____

16. Meter bases shall be installed so that they are not more than ____ not less than ____ from grade level to center of base.

17. The service equipment shall be installed so that top of the top breaker is not more than ____ above floor level.

18. The conductors used to supply ceiling light outlets shall have an insulation rating of ____ C.

19- The minimum size of copper grounding conductor permitted for a UFER (concrete encased electrode) grounding electrode is _____

20. The maximum current that a #14 NMD7 copper cable, which is rated for 90 degrees C. may carry is _____amp.

SECTION THREE   ADDITIONAL QUESTIONS

Write your answers on a separate sheet of paper

Marks

2   I. May the service neutral for a single family dwelling be bare (uninsulated)?
2   2. May flexible cord be used as a substitute for fixed wiring?
2   3. What does the term "readily accessible" mean?
5   4. Calculate the maximum expected expansion in a 12.2 m straight run of 3/4 in. PVC conduit located where the temperature will vary between +115F to -30F.
5   5. What size conduit is required for the following combination of copper conductors; 12 #12TW, 3 #6TW and 3#10TW.
    6. An apartment has 8 suites. The floor area in 4 suites is 45 m² (484 sq.ft.). Each has a 9 kw. range. a 3 kw. water heater, a 4 amp. gas furnace. The remaining 4 suites are 79 m² (850 sq.ft.) each and each has a 12 kw. range, 3 kw. water heater, 4 amp gas furnace and a 5 amp 120 volt garburator. The house load consists of one 4 kw. dryer and 1 kw. of light. Electric service is 120/240 volt, single phase.
        Calculate the following:

5	a	Main service load in amperes.
5	b	Small suite feeder load in amperes.
5	c.	Large suite feeder load in amperes.
5	d.	Minimum number of branch circuit spaces required by cocle for (i) Small suite. (ii) Large suite.
4	e.	Main service conductor size required.
2	f.	Sub-feeder size to small suite.
2	g.	Sub-feeder size to large suite.
2	h.	Main service ground.

10 7.   A single family residence has a floor area of 2000 sq. ft. including breakfast room and laundry room. There are two 1/3 hp. 110 volt motors in the basement hobby workshop. There will be a 12KW range, 3KW water heater, 4KW dryer, an 11 amp. dish washer and a 3 amp. garburator. Air conditioning is with a 1200VA unit. There are 70 plug outlets and 20 light outlets. There is also a 4KW sauna, 8KW electric heating and a 4KW hot tub heater. Calculate the minimum service size required by code. Service is 120/240 single phase.

      8.   Calculate the wiring requirements for a 5 hp. single phase, 230 volt totally enclosed, non-ventilated motor with Class B insulation for a power saw on a construction project.

2	a.	What size wire and conduit is required for this load?
2	b.	What size code fuse is required?
2	c.	What size overload protection is required?
2	d.	What other motor protection does the code require?
2	e.	What is the purpose of the protection required by (d)?
2	f.	Would #8 TW copper be acceptable if run into the motor terminal box.

5	9. a.	What is emergency lighting?
5	b.	Describe fully one method of providing emergency lighting.

5 10. Describe fully how you would install a recessed light fixture.

2 11. Does the code require the nameplate on electrical equipment to remain accessible after instalation?

2	12 a.	What is the minimum service cap height permitted by code.
2	b.	What is the minimum height of service drop wires over a public roadway?

2	13. a.	When using 3-wire loomex cable, would it be permitted to connect both the black and the red wires in this cable to different breakers in the panel but to the same bus?
2	b.	Give a reason for your answer to question (a).

2	14. a.	What is the maximum rating of the overcurrent device for a lighting branch circuit in a residence?
2	b.	What is the maximum rating of the overcurrent device for a branch circuit supplying polarized outlets in the kitchen if the wiring is #12 loomex (copper)?

     15. You are asked to connect a 46 amp. (primary current) manually operated transformer type welder which has a duly cycle of 80% in a non-hazardous area in an existing building of combustible construction in London, Ont. The service is 120/240 volt, single phase.

2	a.	What is the minimum size copper feeder required by code if TW insulation is used?
2	b.	What is the minimum size aluminum feeder required by code if TW insulation is used?
2	c.	If this feeder can be concealed within the building structure, will the code permit it to be NMD7 or NMW-90?

2	16. a.	What is an EYS fitting?
2	b.	Where is it used?

2 17. When wiring barns or stables may NMD7 cable be used?

     18. You are to install the wiring for the lighting and small motors in a machine shop. The service is 120/240 volt, single phase. The building is of frame construction.

2	a.	Would the requirements permit NMS cable to be used for lighting and plug outlet wiring in this building?
2	b.	Could NMS cable be used to supply the 2 hp. lathe and the 2 hp. compressor motors?

4 19. Two circuits are run side by side but in separate conduits. One is #6 R90 copper, the other is #6 TW copper. They are both of equal length and both carry full rated current. Which circuit will have the greatest I R loss?

2 20   When connecting end to end mounted fluorescent light fixtures may R90 wire be used?

3 21. What is the maximum I R loss permitted on an emergency circuit when unit equipment is used?

2 22. Where must the bathroom light switch be placed.

     23. You are asked to replace a 500 watt electric heater with a 1000 watt heating unit on an existing circuit. The circuit, you discover, already supplies 3 other 500 watt heaters, which are on the same thermostat, and another 500 watt heater in the bathroom on its own thermostat. The load is supplied with 90 degree NMD7, #14 copper cable which is connected to a pair of 15 amp. breakers.

2	a.	Would the regulations permit the larger heater to be connected to this circuit?
2	b.	Because the rules permit overcurrent protection for electric heating circuits to be 125% of the fixed load and because 90 degree cable has been used, would it be permitted to replace the 15 amp. breakers with 20 amp. breakers?

2 24. Is a nurses call system a Class II circuit?

2 25. What is the minimum size underground service conduit in a residence?

2 26. May PVC be run in a wall where it is covered with insulation?

2   27. Is it permitted to run two 2-wire loomex cables directly into an electrical space heater?

2   28. How many outlets may be on a 15 amp circuit (not including any appliance outlets)?

2   29. Must the fridge receptacle be split type in every case?

2   30. Is it permitted to cut new holes in the meter base for conduit entry?

4   31. Is a range receptacle required in every single dwelling?

10   32. An existing 30 amp, 240 volt machine is supplied with 2 #10 copper conductors in 3/4 inch conduit. It is intended to install another 20 amp machine and supply it with 2 #12 TW conductors pulled into the same 3/4 inch conduit. Would it be permitted to do this? Give reasons.

33. A building has 2200 sq. ft. of office space and 20,000 sq. ft. of warehouse space. There is also a 3 kw. water heater in the office area. There will be a gas boiler supplied from the warehouse panel for heating. There are three 4 amp motors on circulating pumps and seven 4 amp motors on unit heaters. All motors are 120 volt, single phase. Service is 120/240 single phase. Calculate conductor sizes for:

2   a. The main service.

2   b. The sub to office.

2   c. The sub to warehouse.

2   d. What is the calculated service demand load in amperes?

That, was a tough exam!

<h1 align="center">EXAMINATION ANSWERS</h1>

SECTION ONE ANSWERS

1. T   Rule 2-400(2)
2. T   Rule 26-700(13)
3. F   Rule 10-516 re. to services only. Meters in subfeeders in an apart. build. are not service equipt..Rule 10-204(d) prohibits neutral bonding in these bases.
4. F   Rule 26-702(22),See also page 68.
5. F   Rule 20-202(l) refers. to ``other similar volitile flammable liquids''.
6. F   Rule 68-058(2)
7. T   If in the same structural section, Rule 68-058(4).
8. F   Rule 26-704(9)
9. F   Rule 4-004(10)
10. F   The Building Code requires this.
11. T   Note Rule 2-108 re: Workmanship and cost of material usually controls this.
12. F   Rule 12-1306
13. T   Rule 10-114
14. T   Rule 68-064(2)
15. T   (This is now TW75.) Rule 32-100
16. F   Rule 4-028, 4-030 & 4-034
17. T   Rule 40-008
18. T   Rule 26-704(10)
19. F   Rule 60-400(2) Tele jacks not permitted in bathrooms.
20. T   Rule 12-1404
21. T   Tables based on 30 C ambient plus 30 degree rise.
22. F   Rule 20-004(8) - Class I. Div. I.
23. T   Rule 68-068(7)(c).
24. T   Rule 62-108(3)
25. F   Rule 12-3010(2)
26. F   Rule l2-1402(l)(d).
27. T   Regulation Governing Permits and Fees.
28. T   Rule 32-102(3)
29. T   Rule 2-024.
30. F   Rule 12-102
31. T   Rule 26-704(9)

SECTION TWO

I.   6 times ext. dia.. Rule 12-712(l)

2.   900mm. Rule 12-3016.

3. 80% of br. cir. o.c. prot.Rule 34-018
4. 600mm, 450mm,Rule 12-012(1)&(2), Table 53
5. 15000 volts, Rule 34-204(1)
6. 32 mm, Rule 12-516, page 46.
7. 82.6 in. (2.1 m), Rule 30-318
8. 12, Rule 12-3000
9. Rigid metal conduit or cables approved for hazardous locations, Rule 18-104(1)
10. 7.5 m, Rule 62-114(4).
11. 6mm, Rule 12-3018
12. #6, Rule 68-402(1)
13. 4800, page 101
14. 1 m, Rule 6-112(3)
15. 900 mm, Rule 2-322
16. 1850mm 1650 mm page 20
17. 1.7 m, 67 inches, page 28.
18. 90^0, Rule 30-412(1).
19. #4, Rule 10-702(2)(a)
20. 15, Table 2

SECTION THREE

I.   Yes, Rule 6-308

2.   No, Rule 4-010(3)

3.   Page 45 in the Code book.

4.   First change F to C. then, 12.2 m x 62.8°C x 0.052 mm = 39.8 mm.

5.   1 1/4 inch, Tables 8, 9, 10.

6.   a.   196.9 amps.
      b.   42.7 amps
      c.   48.9 amps
      d.   (i) 8 circuits, Rule 8-l08(3)(a)
          (ii) 8 circuits
      e.   3/0 RW75, 250 MCM RW75
      f.   #6 RW75,#6 R90
      g.   #6 RW75,#6 R90
      h.   #3 copper, aluminum not permitted - Rule 10-802

7.   126 amps.

8.   a.   #8 R90 copper in 3/4in. conduit direct to motor. Length between motor starter and motor termination must be mini-

mum 1.2 m (47.2 in.), see Rule 28-104.
- b. 90 amp.
- c. 35 amp.
- d. Magnetic starter, Rule 28-400
- e. Rule 28-400, page 603 of Code.
- f. No, Rule 28-104

9. a. These are lights which automatically give minimum illumination in specific areas when the normal power supply to the building fails.
   - b. An emergency lighting system may consist of two or more lamps directed to light specific areas of a room. The power supply for these lights consists of a metal enclosure which contains batteries, charging equipment, control equipment and a relay which is kept connected to the building power supply. In the event of normal power failure this relay becomes de-energized, falling into the open position, thereby closing a set of contacts which complete the circuit through the lamps and batteries. In this way the emergency lighting system is operating the moment after normal power failure. Rule 46-300.

10. Page 52
11. Yes, Rule 2-118
12. a. 4.5 m plus 300 mm = 4800, Rules 6-112(2) & 6-116(b)
    - b. 6 m, page 14
13. a. no
    - b. The red and black wire must be connected to give 240 volts in order to balance the load on the neutral. If this is not done the neutral conductor would have to carry the combined load of the two 120 volt circuits.
14. a. 15 amp, page 43
    - c. 15 amp, page 43
15. a. #8, Rule 42-006(1)(a)
    - b. #6, Rule 42-006(1)(a)
    - c. Yes, Rule 12-504
16. a. A conduit seal used in hazardous locations.
    - b. Rule 18-106(1)&(3)

17. No, Type NMW or NMWU is required, Rule 22-204
18. a. Yes, Rule 12-504.
    - b. Yes, Rule 28-100
19. The R90 conductors because they carry more current, Table 2
20. Yes, Rule 30-312(2)(a) & (b) Note, must have 600 volt rating.
21. 5% , Rule 46-306
22. 1 m (39.4 in.) from the tub or shower, Rule 26-702(12)
23. a. No, Rule 62-114(6)
    - b. Yes, page 100
24. No, Rule 16-010

25. 2 in. Building Code Rule 9.34.4.5.(1)(a)

26. No, Rule 12-1102(2), page 21

27. Yes

28. 12, Rule 12-3000(1)

29. No, Rule 26-704(2), page 71

30. Not if base is outdoors. All holes to be below any live parts inside base.

31. Yes, Rule 26-746(5) unless built-in gas or electric range is installed, Rule 26-746(13)

32. No, conduit fill is okay, Tables 8,9 and 10 but conductor ampacity is only 80%. Rule 4-004(1)(c)

33. a. #3 R90 copper
       #1 R90 aluminum
    - b. #6 R75, #4 R75
    - c. #6 R75, #4 R75
    - d. 103.4 amperes

OK we did it! Now what is our percentage mark?

To find out do this;
        a. Add all the marks for all the **correct** answers•
        b. Then add all the marks for **incorrect**" True or False" answers only.

Your formula should look like this;

$$\frac{a \text{ less } b}{\text{Total points possible}} \times 100 = \underline{\hspace{1.5cm}} \% = \frac{a - b}{266} \times 100 = \underline{\hspace{1.5cm}} \%$$

Minimum circuits required 35 43